THE MAGNIFICENT SKELETON

BONE ANATOMY AND PHYSIOLOGY FOR ASPIRING YOUNG DOCTORS

JOHN WINTERS

CONTENTS

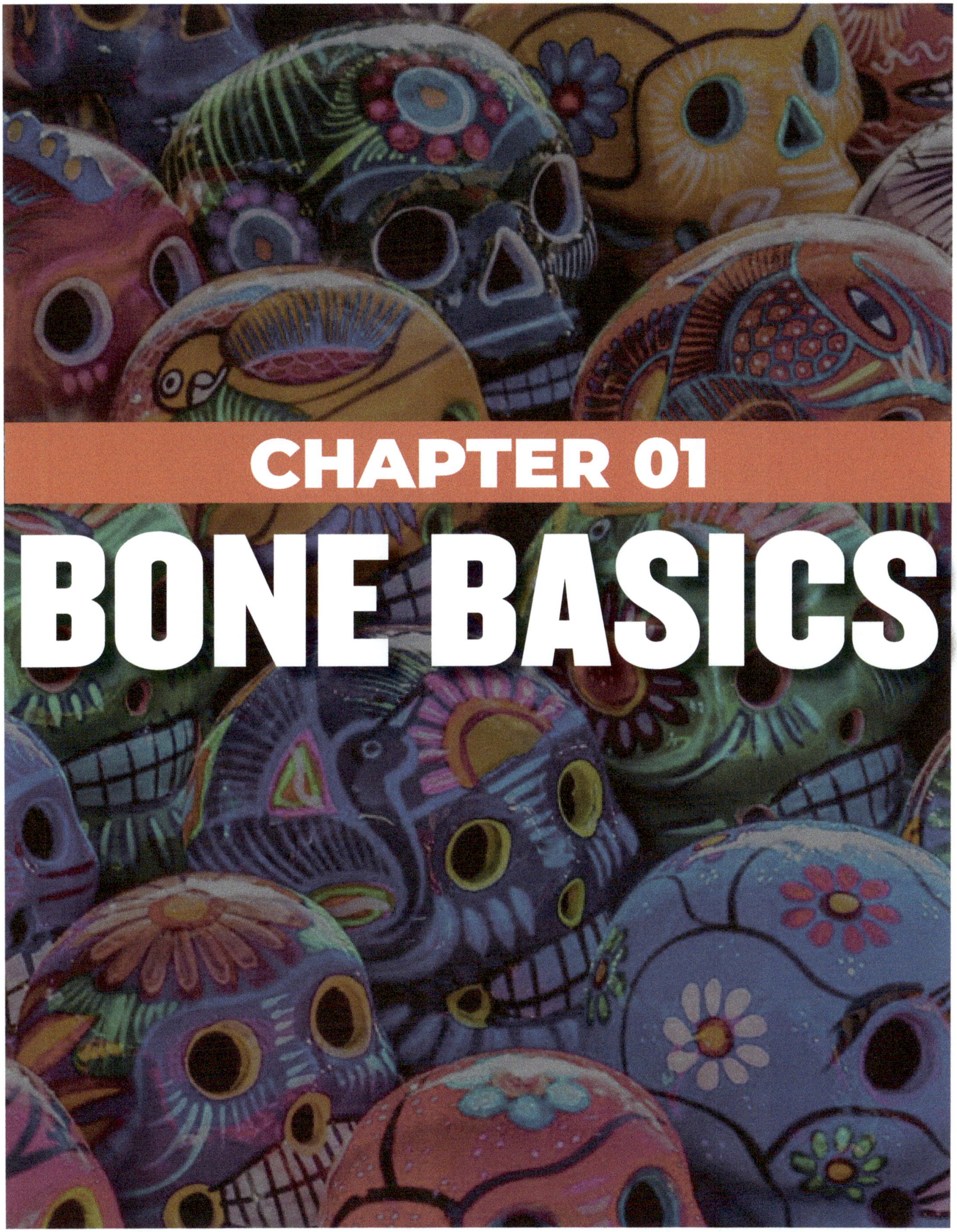

CHAPTER 01

BONE BASICS

WHAT IS A BONE?

Bones often don't get the recognition they deserve, as our hair, skin, and nails tend to steal the spotlight. However, bones are incredibly important as they form the foundation of our skeleton.

Various unique bone structures come together to perform essential roles: they give us, nourishment, stability, shape, and help enable us to move.

Let's dive into some bone basics.

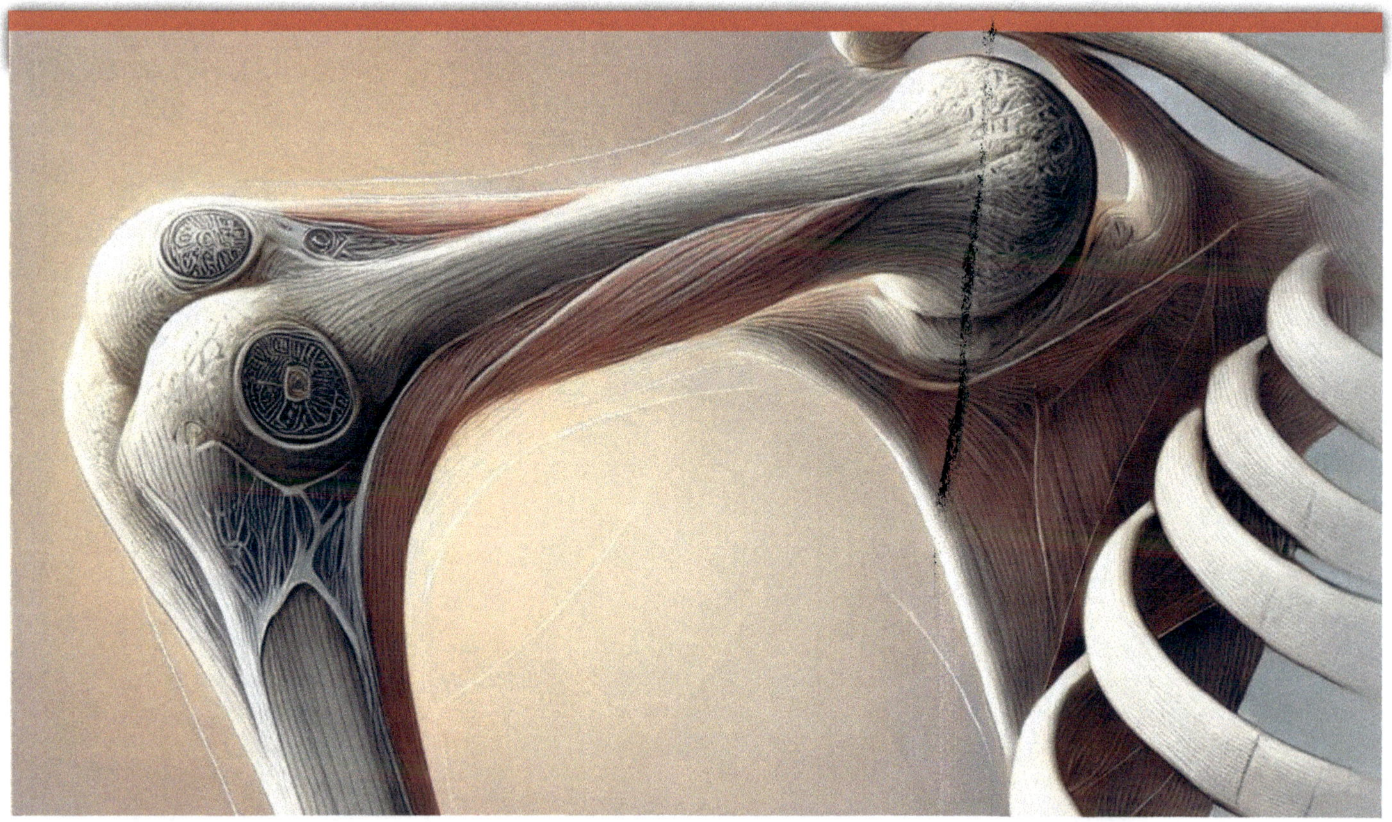

HOW MUCH DO BONES WEIGH?

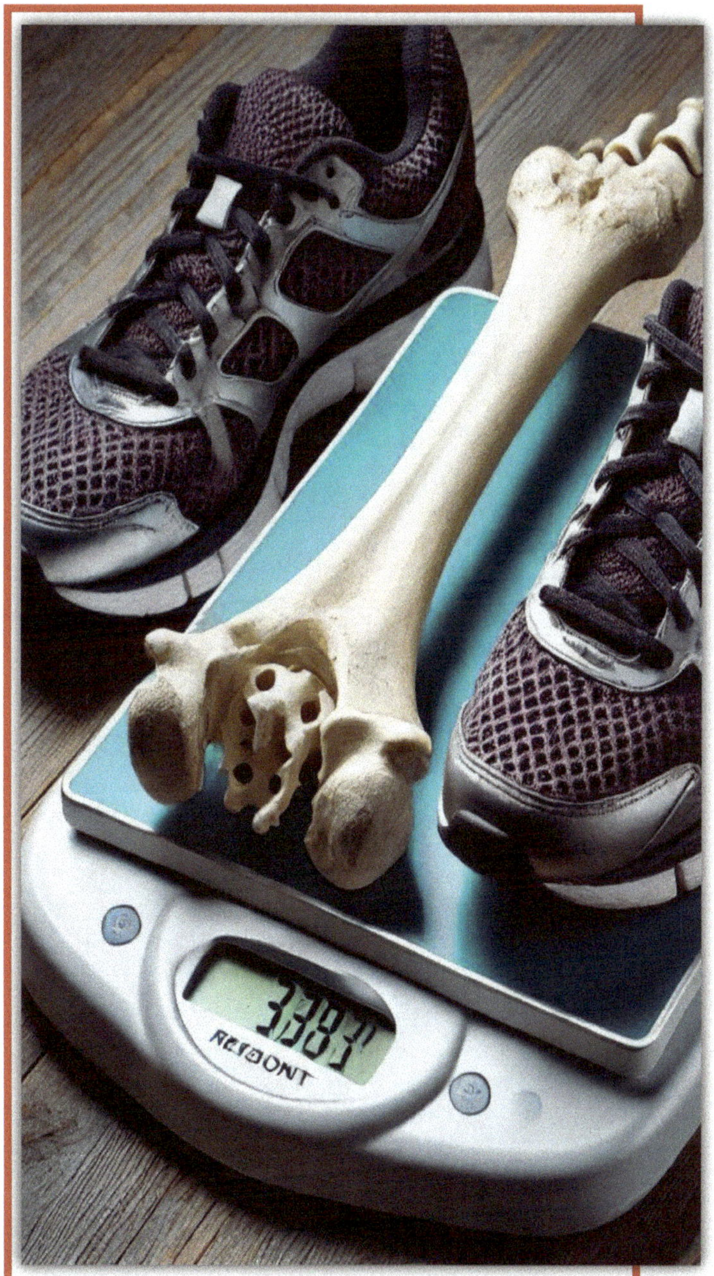

- **Small Bones:** Bones like those in your fingers or toes (phalanges) weigh just a few grams each.

- **Medium Bones:** Bones like the ribs or vertebrae can weigh anywhere from 100 to 200 grams.

- **Large Bones:** The femur (thigh bone), which is the longest and heaviest bone in the human body, can weigh about ***500 to 700*** grams or ***1.5 pounds***. That's about the weight of a large grapefruit or a pair of running shoes.

Keep in mind that these weights can vary based on factors like age, gender, and overall body size. The skeleton as a whole makes up about **15%** of a person's total body weight.

WHAT ARE BONES MADE OF?

Bones are made of:

Water - 50%
Minerals - 26%
Proteins- 20%
Fats - 4%

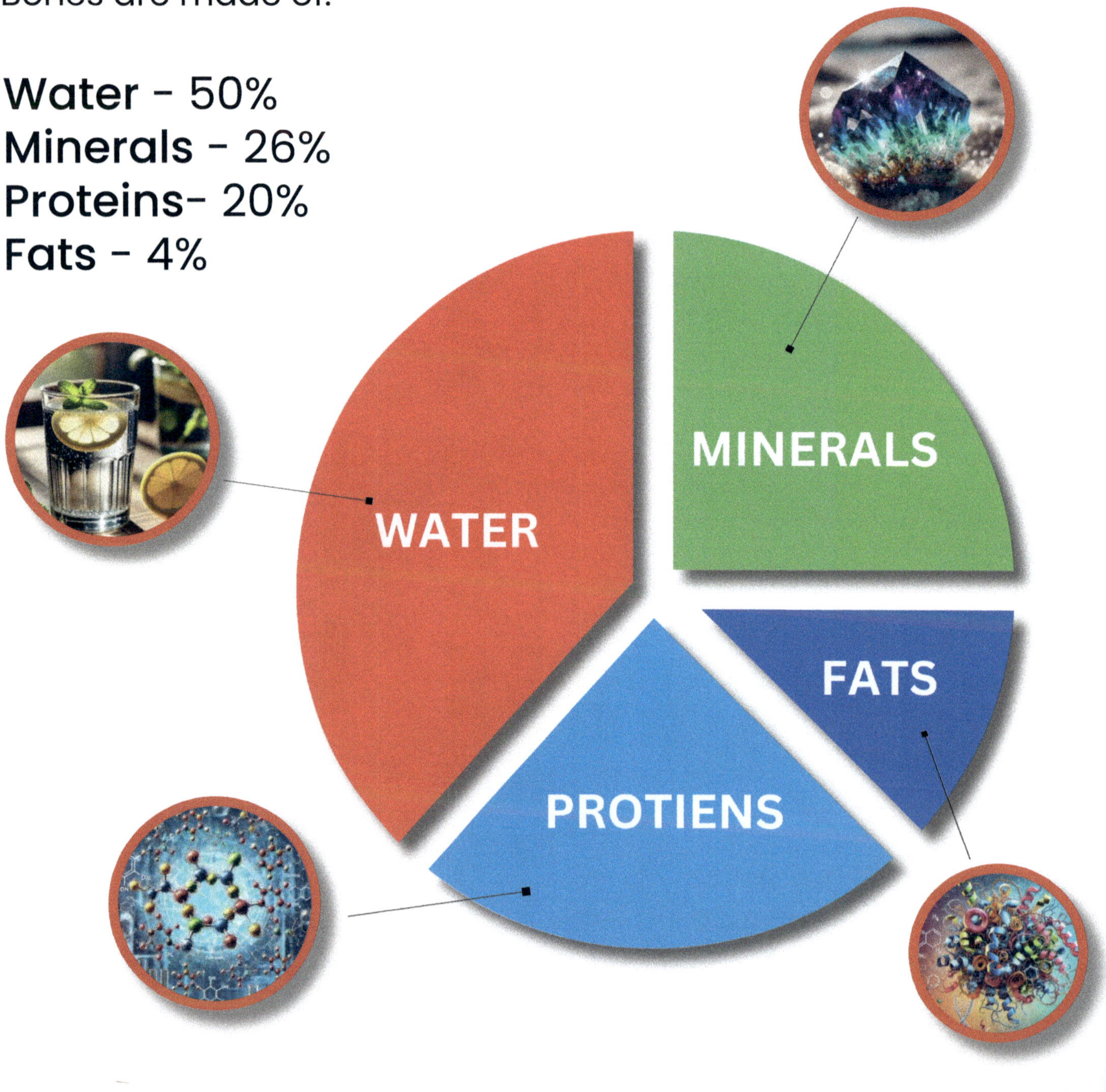

HOW MANY BONES ARE THERE?

The human body typically has *206* bones.

However, newborn infants start with a higher count of around *270* bones. As they grow, some of these bones gradually fuse together to form single bones, which explains the reduced number in adults.

Let's count the number of bones in each area:

- Skull (22 bones)
- Vertebral Column- spine (33 bones)
- Ribcage (25 bones)
- Arms and Shoulders (64 bones)
- Legs and Pelvis (62 bones)

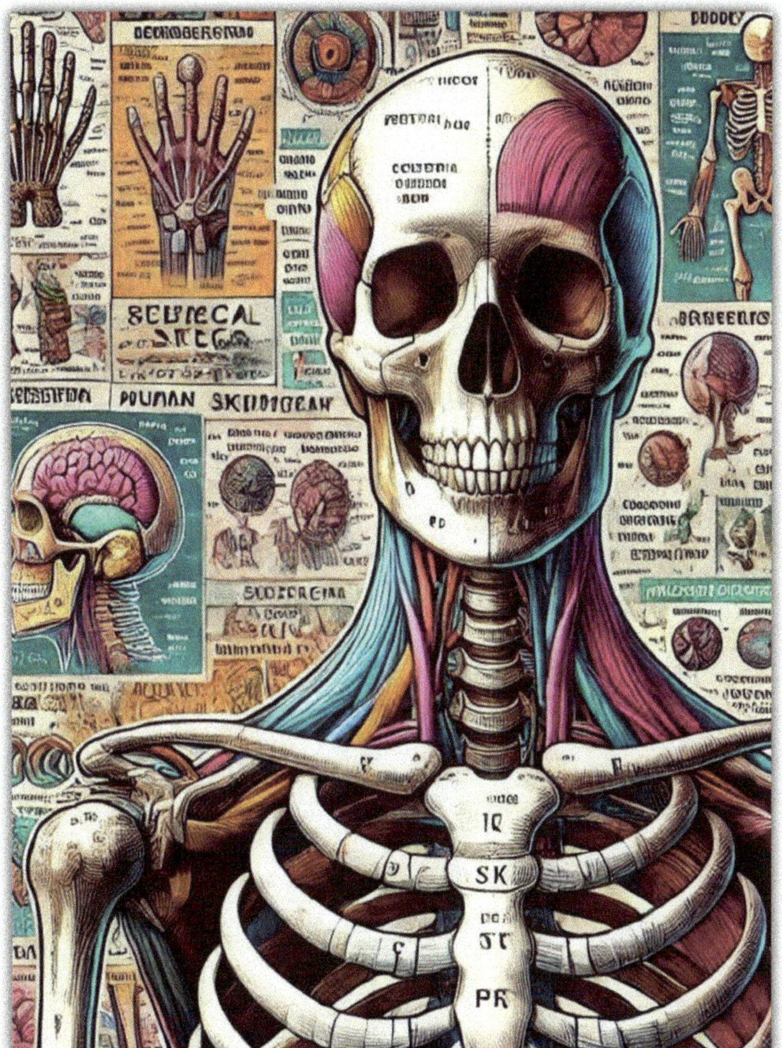

HOW STRONG ARE BONES?

Bones are remarkably strong structures, often compared to the strength of materials such as steel and concrete. Here are some key points about the strength of bones:

Tensile Strength: Bones can withstand considerable tension (stretching forces). They have a tensile strength like cast iron but are much lighter.

Compressive Strength: Bones are also excellent at with standing compression (pressing forces). The compressive strength of bone is comparable to that of reinforced concrete.

FUN FACTS ABOUT BONES

FUN FACT # 01

Did you know that the **smallest bone** in the human body is the **stapes** bone in the middle ear? It's just about **3 millimeters** long and is shaped like a tiny stirrup. Despite its small size, it plays a crucial role in helping us hear by transmitting sound vibrations to the inner ear.

FUN FACT # 02

The **hyoid bone** is the only bone in the human body that does not directly connect with any other bones. It's located in the throat, and its primary function is to support the tongue and help with speech and swallowing. This unique bone is anchored by muscles and ligaments, making it a vital yet isolated component of the skeletal system.

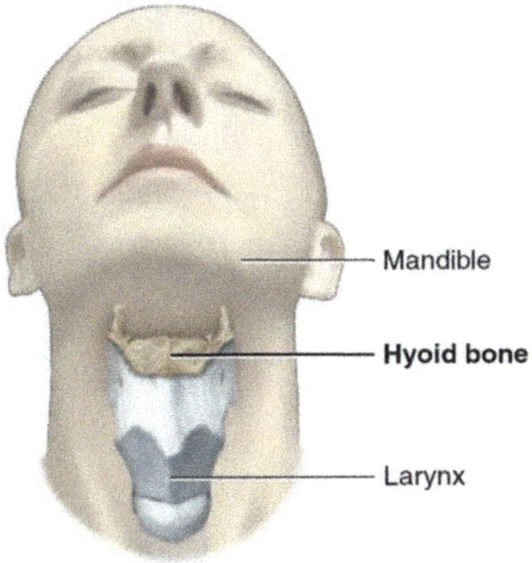

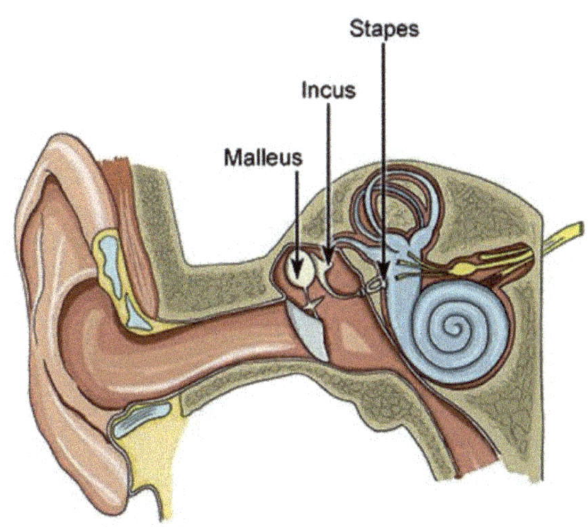

CHAPTER 02

TYPES OF BONES

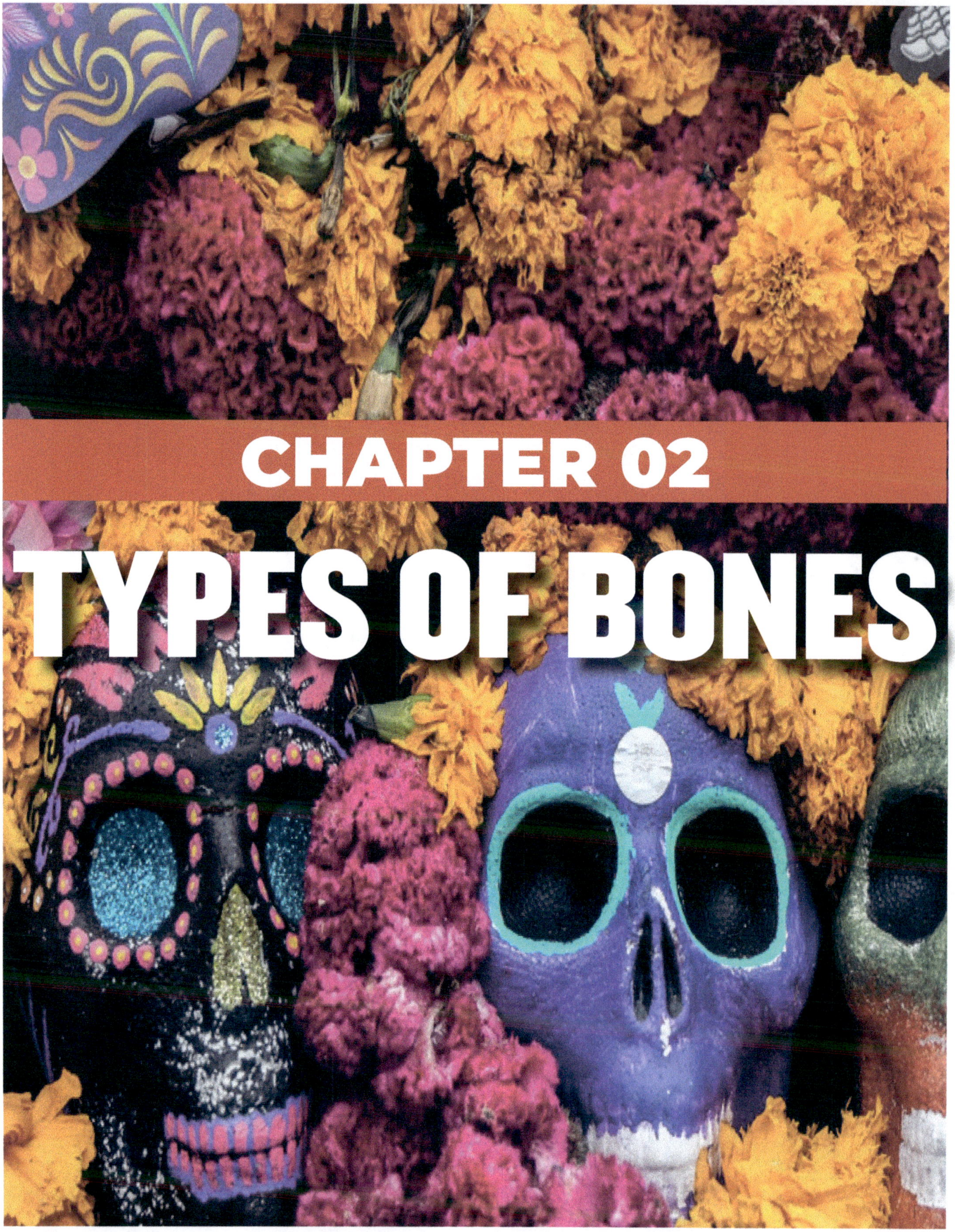

BONE SHAPES

There are many different types of bones in the human body. They have different shapes, and each has a specific purpose. Let's review the various designs.

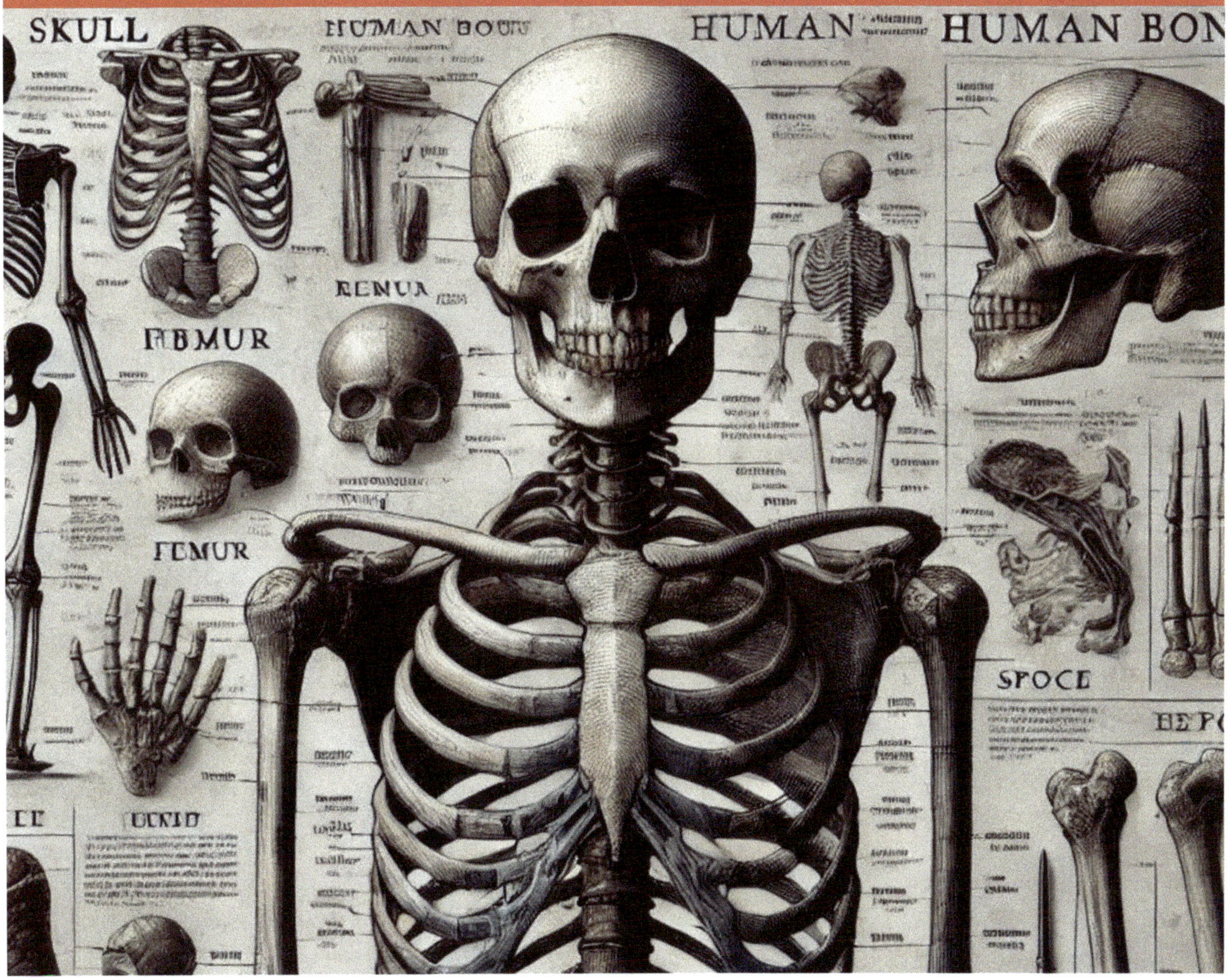

LONG BONES

Long Bones
These bones are longer than they are wide.

Purpose: Long bones provide support and help in movement.

Examples: Femur (thigh bone), humerus (upper arm bone), tibia (shin bone)

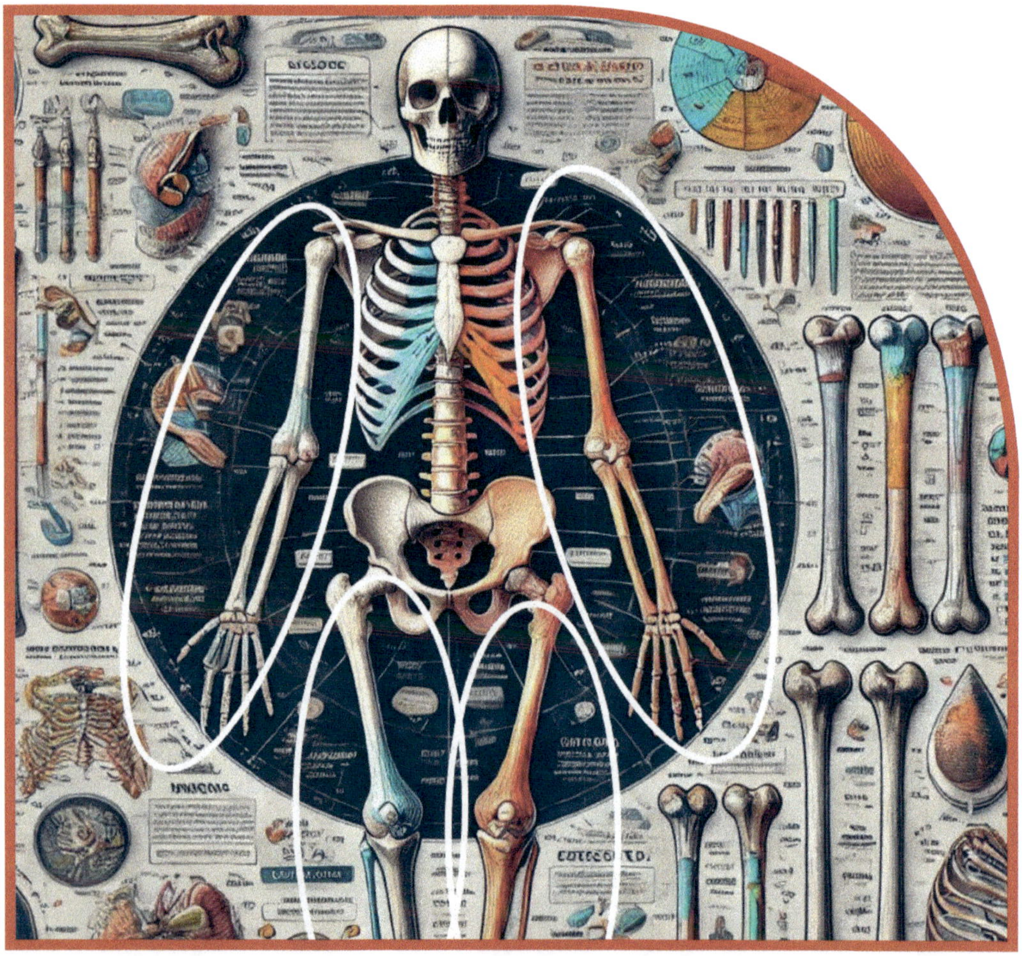

SHORT BONES

Short Bones: These bones are roughly as long as they are wide.

Purpose: Short bones provide stability and help with movement and grabbing things.

Examples: Carpals (wrist bones), and tarsals (ankle bones).

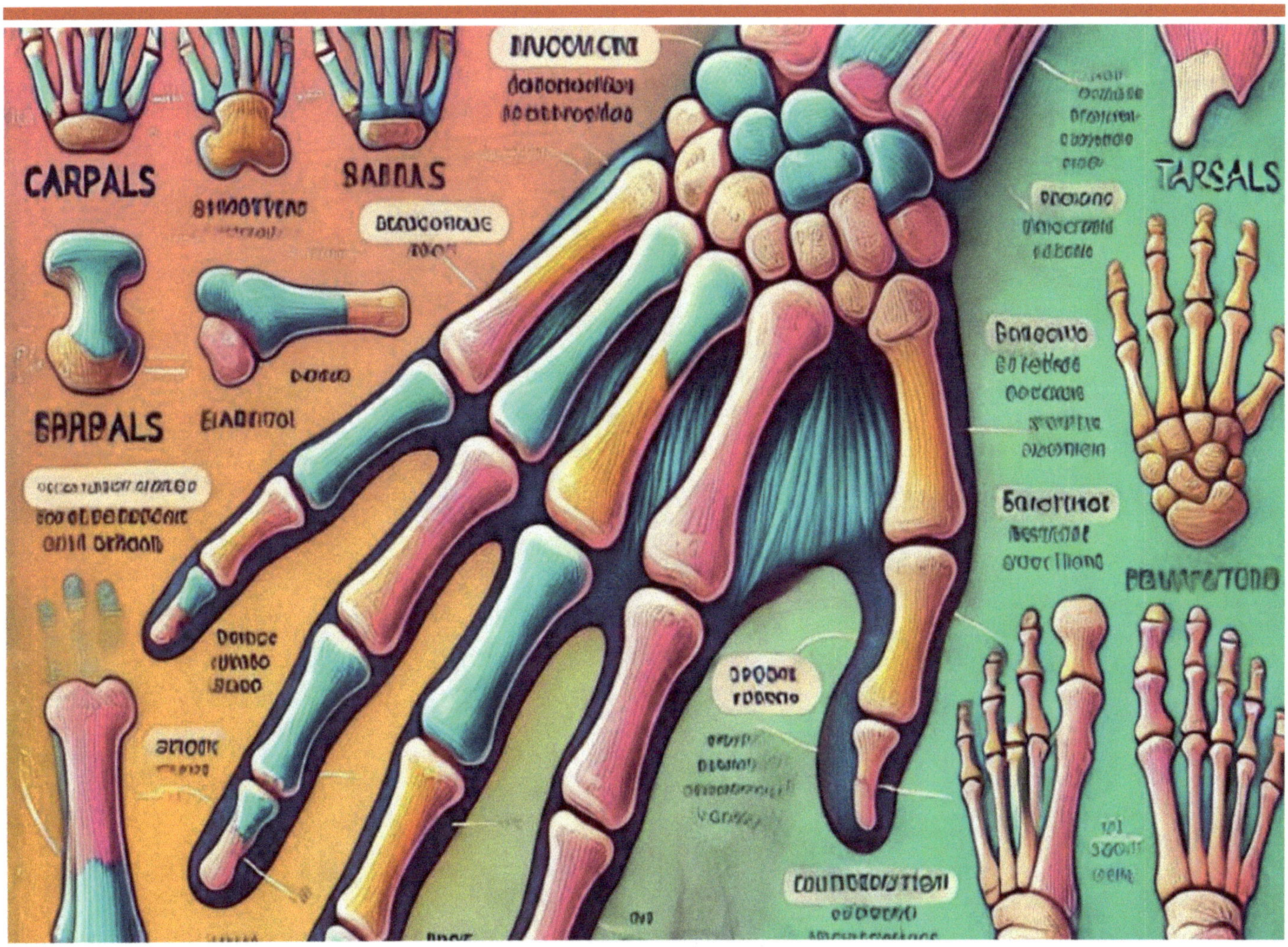

FLAT BONES

Flat Bones: These bones have a thin, flattened shape.

Purpose: Provide protection for underlying organs or broad surfaces for muscle attachment.

Examples: Skull bones, ribs, sternum (breastbone), scapulae (shoulder blades).

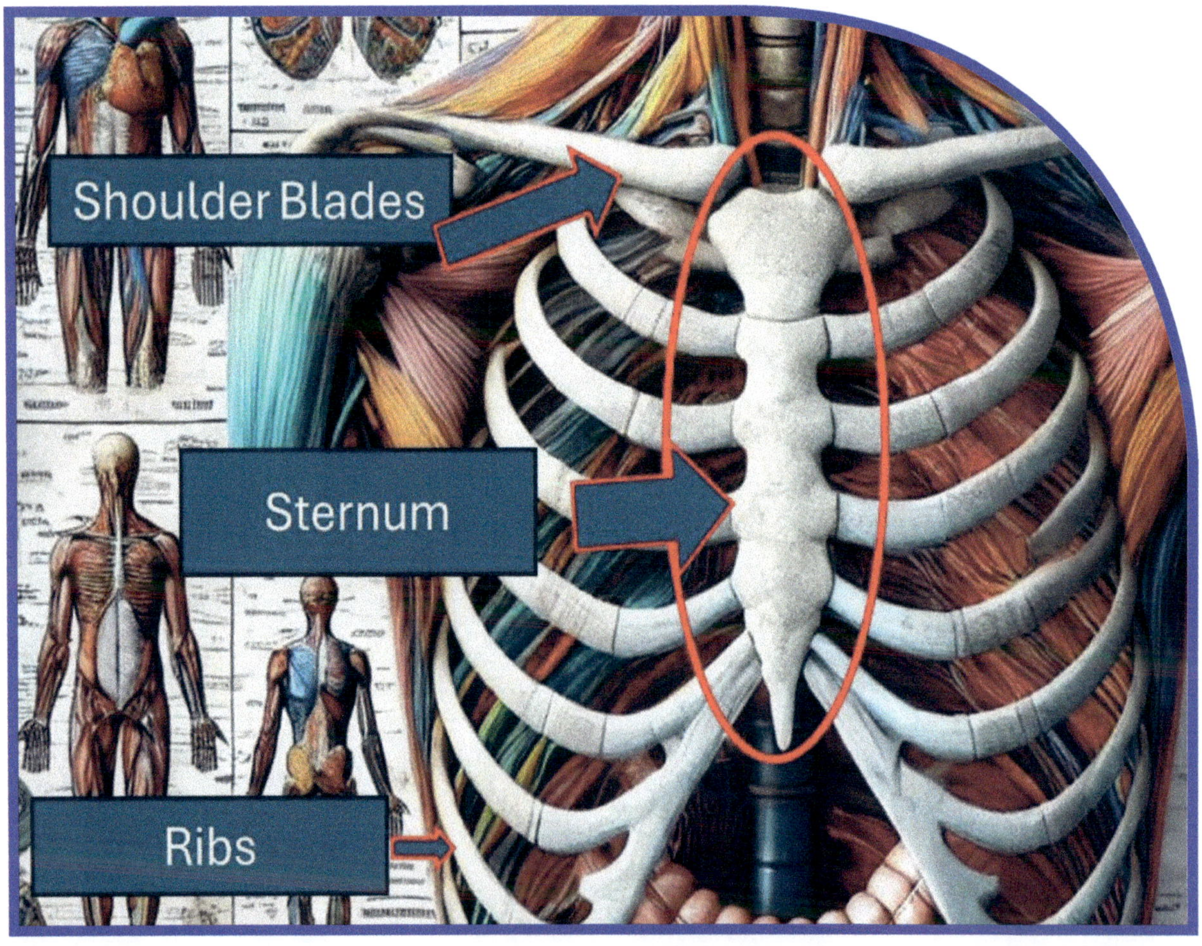

SESAMOID BONES

Sesamoid Bones: These small, pea-shaped, round bones are embedded within tendons.

Purpose: These bones protect by reducing friction and modifying pressure to specific areas of the hands and feet.

Examples: Patella (kneecap), small bones found in tendons of the hands and feet.

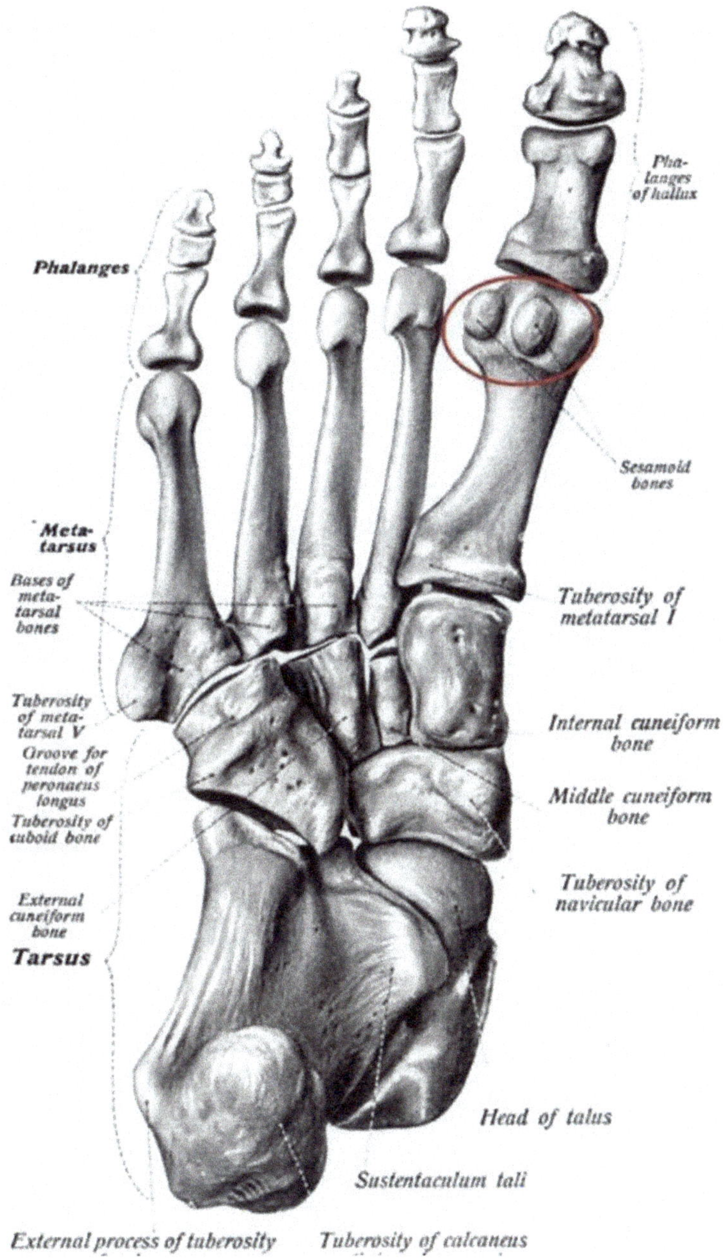

IRREGULAR BONES

Irregular Bones: These bones have complex shapes that do not fit into the other categories.

Purpose: Irregular bones serve various roles, such as protection, support, stability and movement.

Examples: Vertebrae (spine bones), sacrum (pelvis), mandible (jawbone).

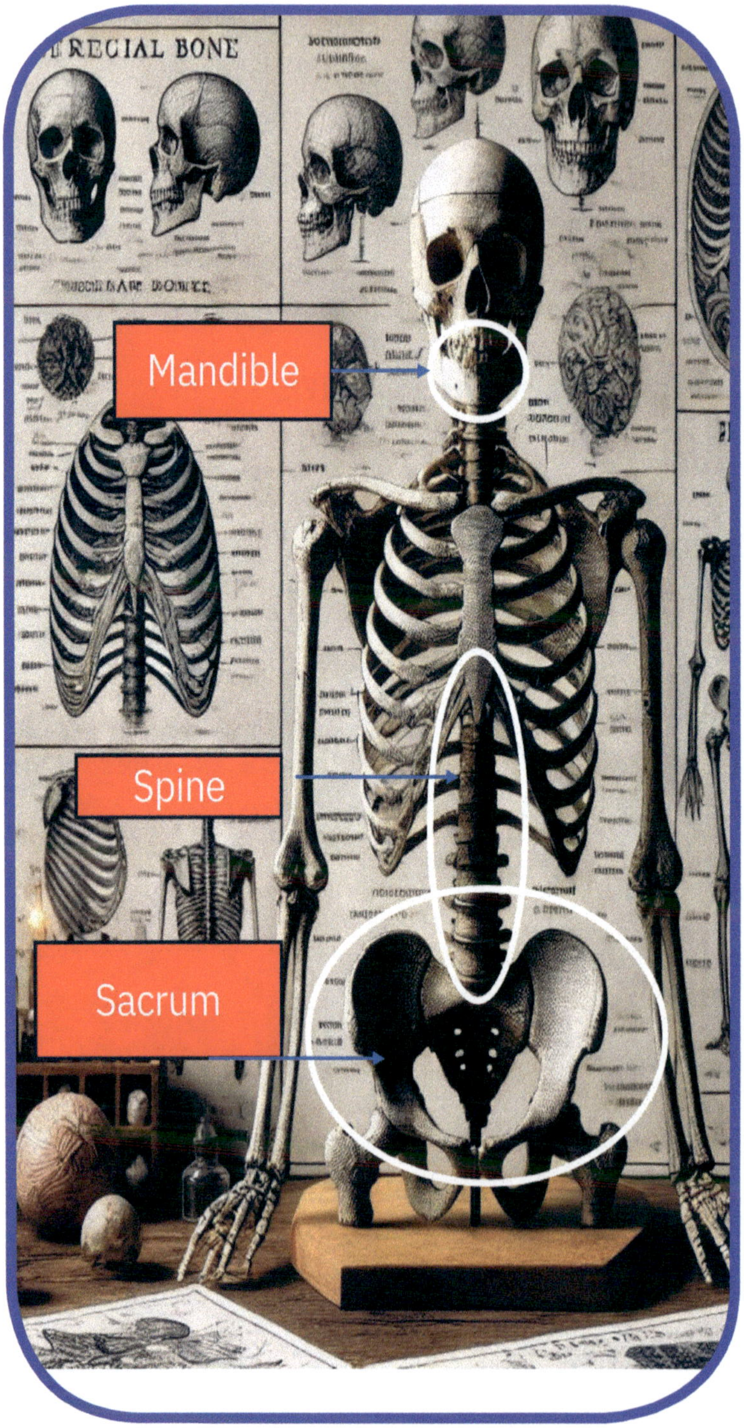

THE TWO SKELETONS

The human skeleton is divided into two main categories. They are called the ***Axial Skeleton*** and the ***Appendicular Skeleton.***

The **axial skeleton** is the central core of the human skeleton, providing support and protection for the brain, spinal cord, and vital organs in the center of the body.

The **appendicular skeleton** is the portion of the skeleton that includes the bones of the limbs and the pelvis region. This skeleton attach all the limbs to the axial skeleton. It's essential for movement and interaction with the environment.

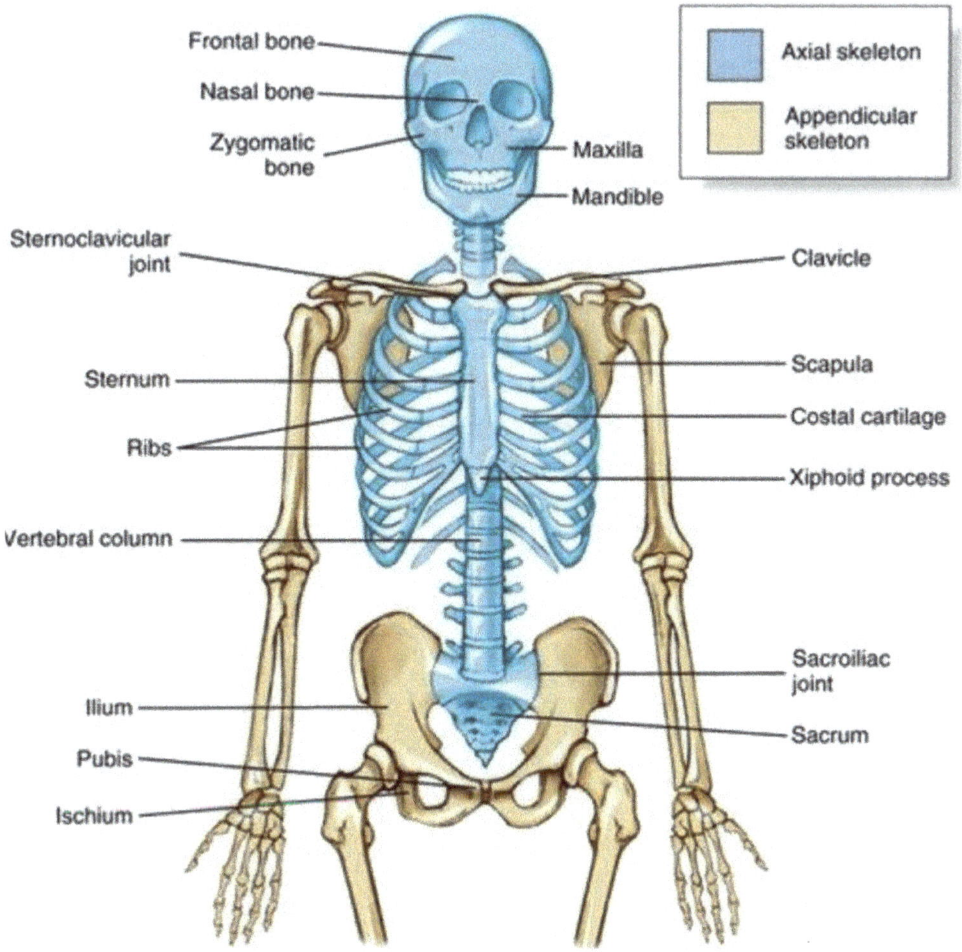

BONE LIFE FUN FACT

Bones can last for **thousands**, even **millions**, of years under the right conditions. Here's why:

1. Environment: Dry, arid climates and places with stable temperatures preserve bones better. Conversely, wet and acidic environments can cause bones to decay faster.

2. Burial Conditions: Bones buried quickly and deeply in sediment are protected from the elements, scavengers, and environmental changes, leading to better preservation.

3. Mineralization: Over time, minerals from the surrounding environment can replace the organic materials in bones, turning them into fossils. This process, called permineralization, can make bones last millions of years.

4. Soil Chemistry: Alkaline soils preserve bones better than acidic soils, which can degrade bones more rapidly.

CHAPTER 03

PURPOSE OF BONES

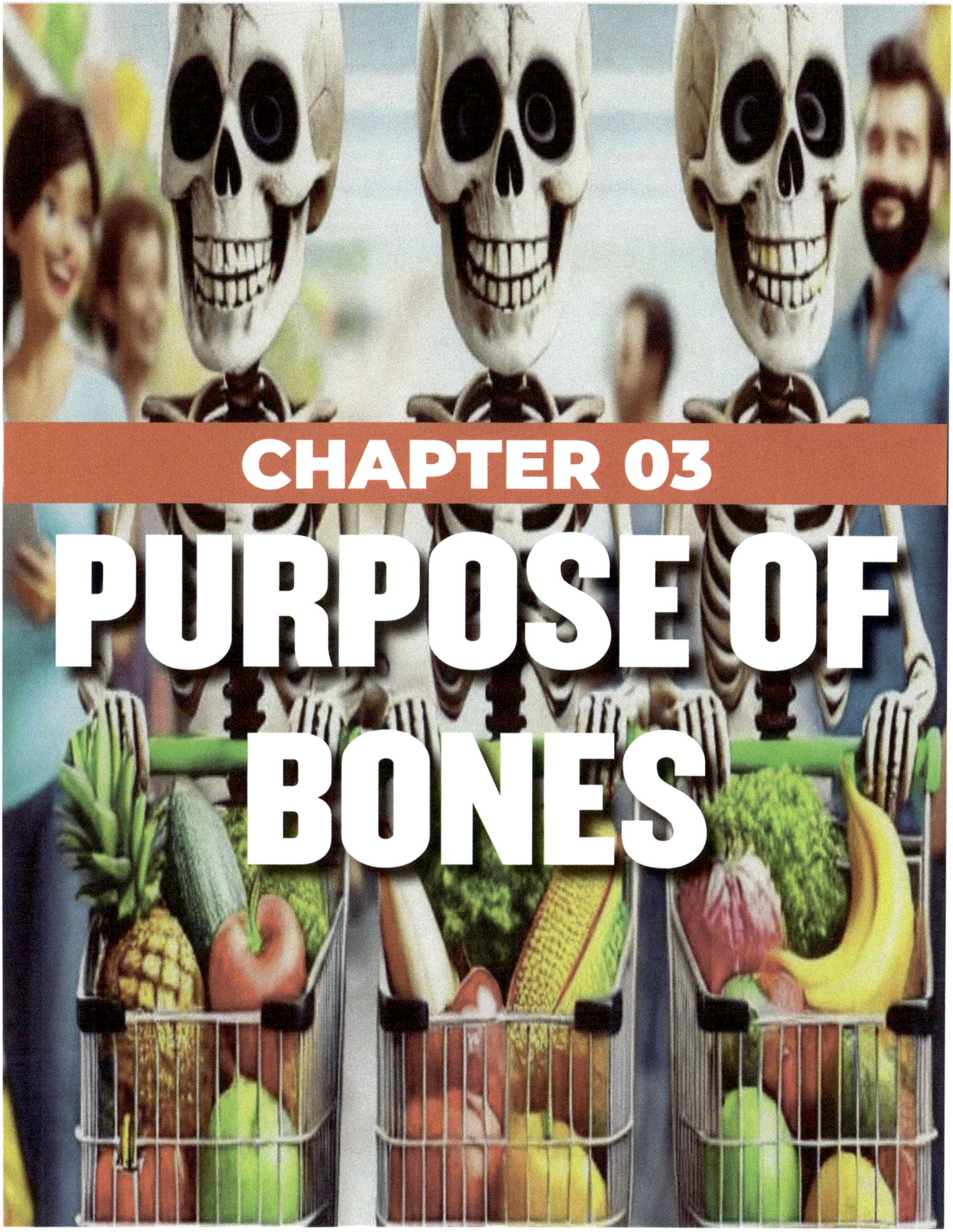

WHY STUDY BONES?

The study of bones is known as osteology. Osteologists examine the structure, function, and diseases of bones. This field is crucial in various areas, including:

- **Medicine:** Understanding bone health, diagnosing and treating bone-related conditions.

- **Archaeology:** Analyzing ancient human civilizations to learn about past populations, their health, diet, and lifestyles.

- **Forensic Science:** Determining causes of death in legal investigations.

Let's review other reasons why bones are so important.

BONES CREATE STRUCTURE

Without bones, we would appear as blobs sliding around. We might even resemble jellyfish, worms, or sea anemones. These creatures have no bones or skeletons.

FUN FACTS

Creatures with skeletons are called "Vertebrates."
Creatures without skeletons are called "Invertebrates."

BONES HELP US MOVE

Our bones allow us to move, jump, lift things, and run. They allow us to do many of the physical activities that we like doing.

FUN FACTS

Our muscles and skeleton work together to give us the ability to move. This combined effort forms the Musculoskeletal System of the body.

BONES PROVIDE PROTECTION

Our bones do more than allow us to run and jump, they also protect vital and sensitive organs of our body like the heart, lungs, brain, spinal cord, and spleen. Our skeleton acts as a protective shield around our insides to keep them from harm.

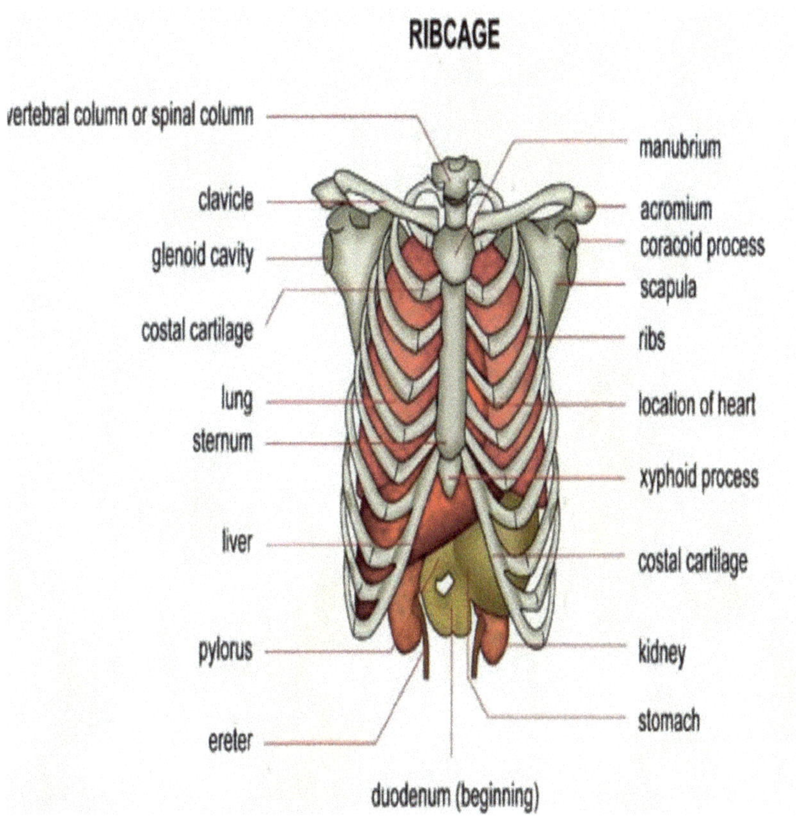

FUN FACTS

The rib cage not only protects vital organs but also plays a key role in breathing by expanding and contracting

BONES CREATE BLOOD

Our bones also create blood for our bodies. Blood is very important because it carries the oxygen that allows us to breathe, help fight infections and heal wounds. The forming of blood takes place in a part of the bone called the bone marrow, which is a soft, spongy tissue found in the hollow centers of certain bones.

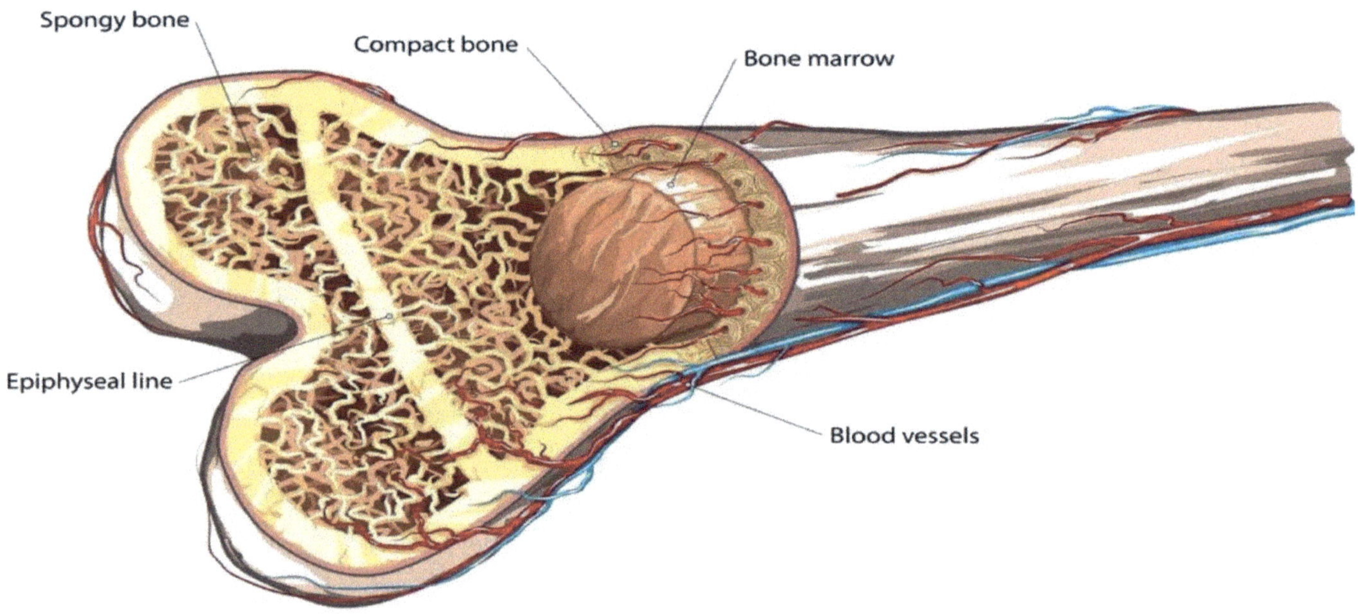

FUN FACTS

Blood is made up of three components. They are called Platelets, White Blood Cells, and Red Blood Cells. They're all created inside the thigh bone.

BONES STORE NUTRIENTS

Our bones also store important minerals like calcium, magnesium, and phosphorous. In addition, our bones can store energy in the form of fat. When the body needs these nutrients, the bones will release them at the right time.

FUN FACTS

Think of this like a small refrigerator in your body. When the body is low on these nutrients, the bones provide them.

KNOWLEDGE CHECK

Choose the best answer

1. What is the study of bones called? ○
2. Without bones, what might we look like? ○
3. The rib cage provides what? ○
4. Our bones create something very important. What is it? ○
5. What do our bones store? ○

- ○ Blobs, jellyfish, anemones
- ○ Osteology
- ○ Protection
- ○ Blood
- ○ Minerals

CHAPTER 04

HOW BONES ARE MADE

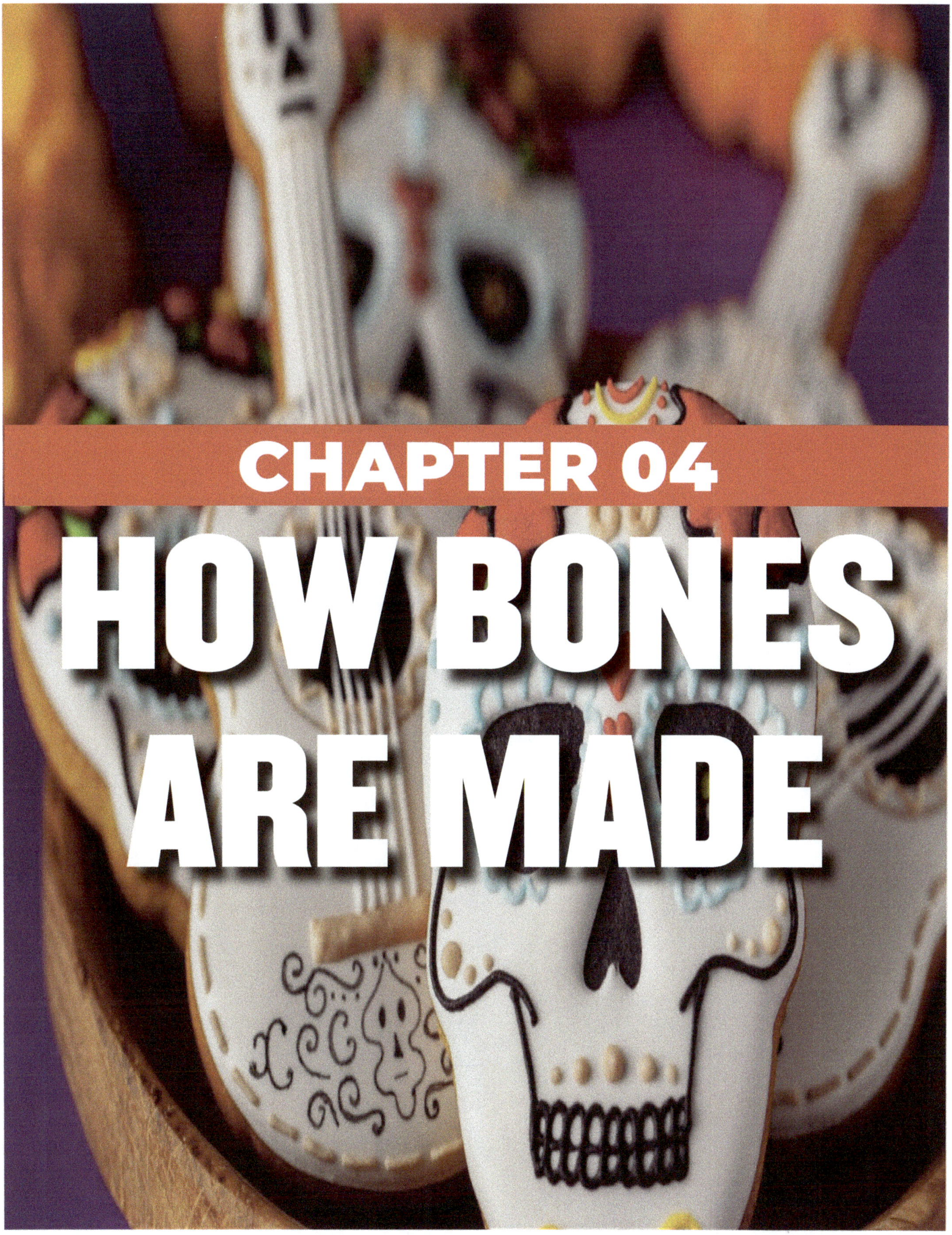

THE TRANSFORMATION

When a baby is very young, bones begin to form from a small component called a **Cartilage**. As the months pass, the cartilage will slowly start to change and transform into bone tissue. This transformation takes place in small steps over time and continues from childbirth well into the teenage years. The process is called **Ossification**.

Think of the cartilage transforming into a bone like a caterpillar that transforms into a butterfly.

HOW THE TRANSFORMATION (OSSIFICATION) TAKES PLACE

1. Cartilage Model: A cartilage model of the bone forms first.

2. Bone Collar Formation: Osteoblasts create a bone collar around the cartilage model.

3. Primary Ossification Center: Blood vessels invade the center of the cartilage model, bringing osteoblasts and osteoclasts (cells that break down bone tissue). The cartilage is replaced by bone tissue.

4. Secondary Ossification Centers: These form at the ends of long bones (epiphyses). The process is similar to the primary ossification center.

5. Growth Plates: Between the primary and secondary ossification centers, the growth plates (epiphyseal plates) remain until adulthood, allowing bones to lengthen.

OSSIFICATION

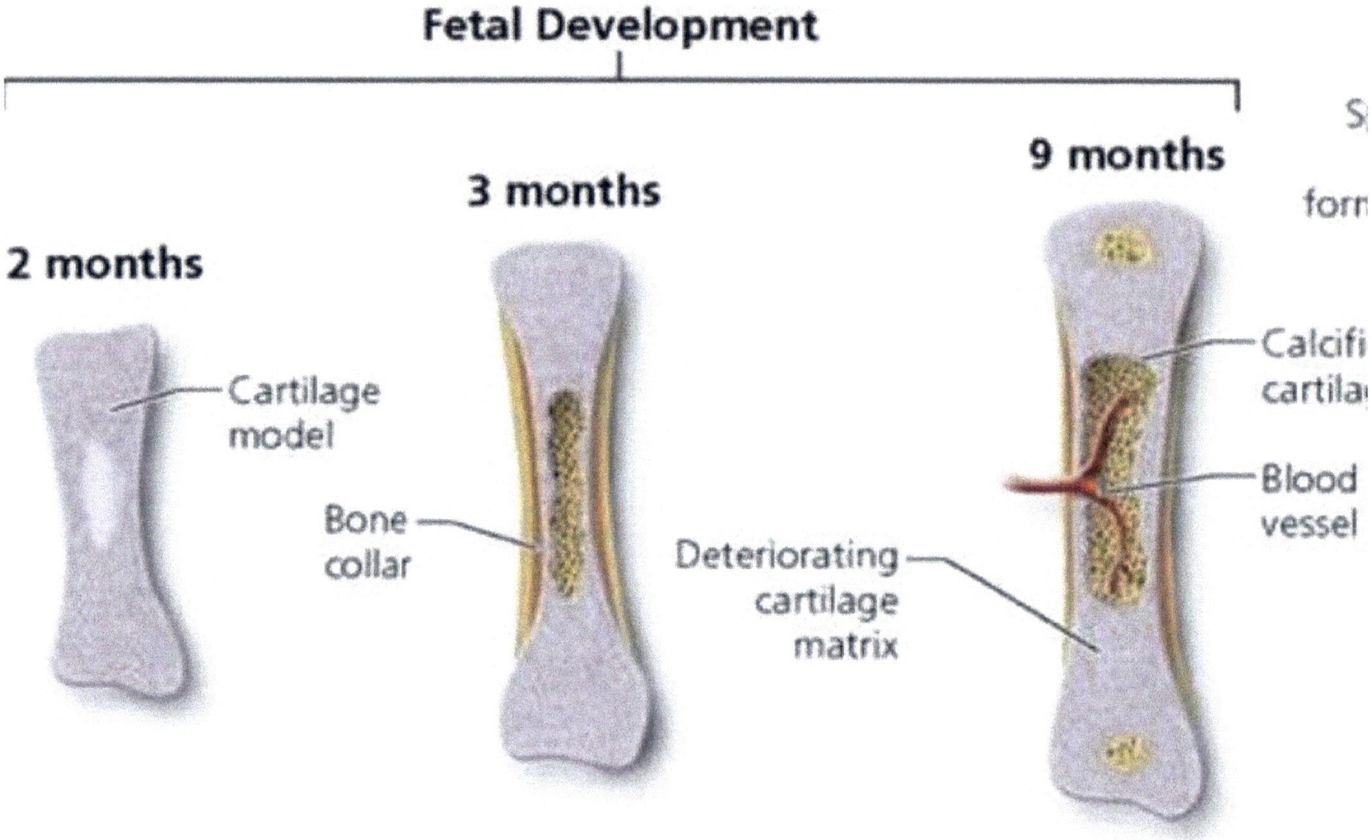

Step 1: A cartilaginous model of the future bone forms.

→

Step 2: Osteoblasts form a collar of bone around the shaft of the model.

→

Step 3: The shaft of t cartilage model begi to hollow out, and spongy bone fills the space. Blood vessels continue to penetrat the area, and the region of bone formation expands.

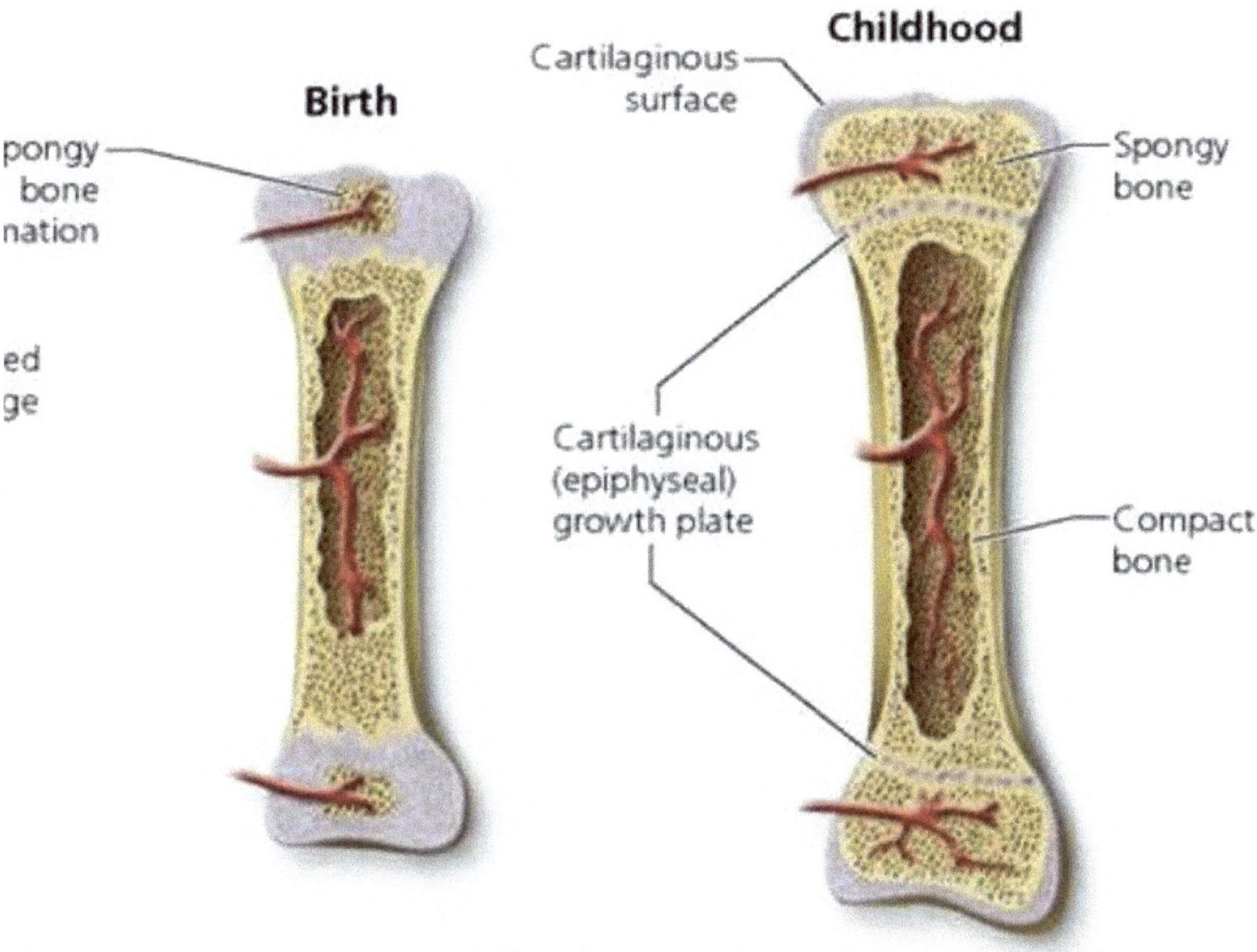

Step 4: Secondary centers of bone formation develop in the ends of the bone.

Step 5: Cartilage remains only on the surfaces that rub against other bones and in the cartilage growth plates.

TRANSFORMATION HELPERS

Two small helper cells assist in the bone transformation process. The first helper is the one that will change the cartilage into bone. They are called **"Osteoblasts."** Their role is to build the bone.

Sometimes bone parts gets damaged during the building process and needs to be removed. This removal is done by small component cells called **"Osteoclasts."**

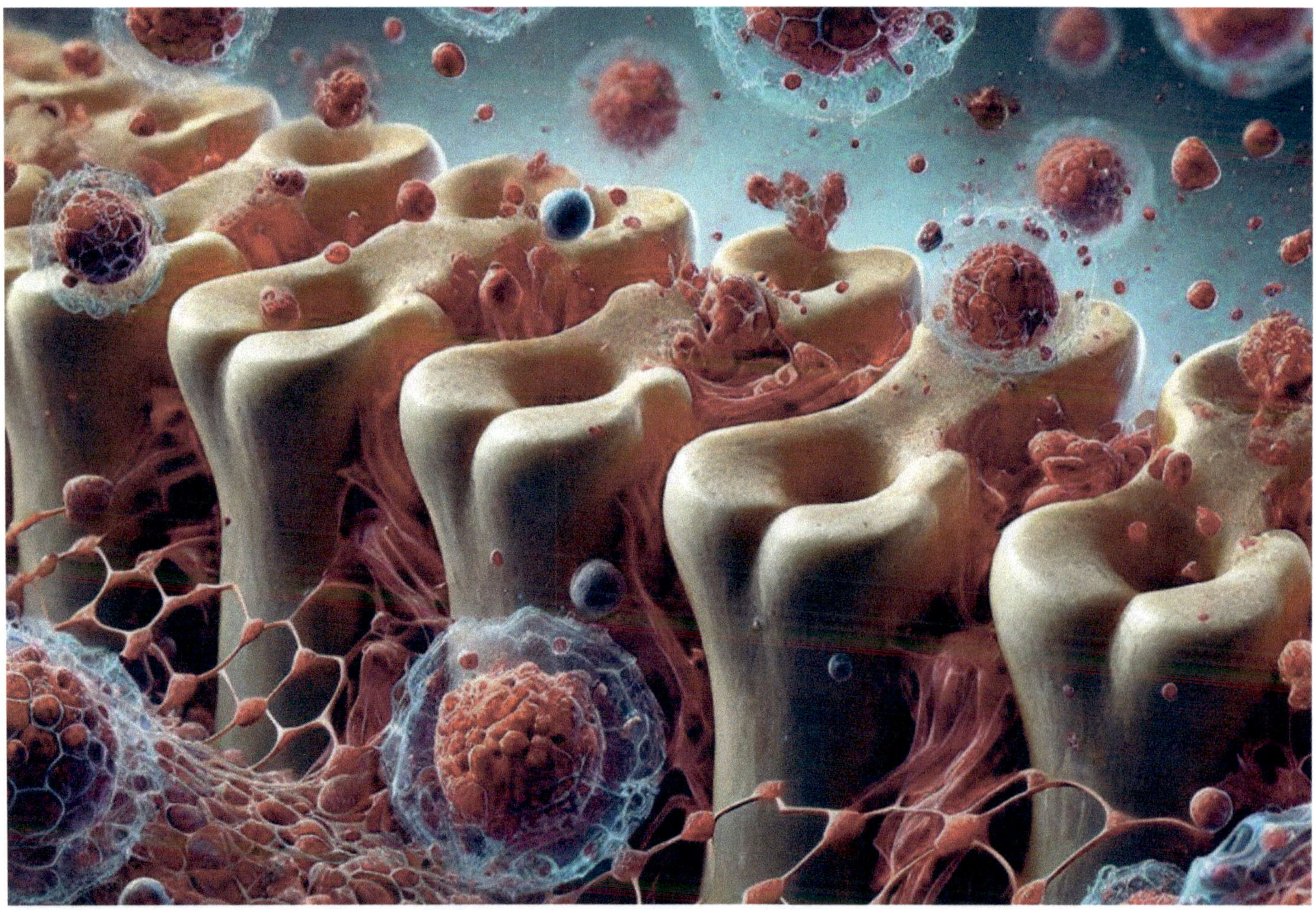

KNOWLEDGE CHECK

Determine if the statement is true or false.

1. Bones start off as a cartilage model. — **True or False**
2. The process of transforming from a cartilage into bone is called ossification. — **True or False**
3. Bones will continue to grow into the teenage years. — **True or False**
4. Osteoblast are cells that help build bones. — **True or False**

CHAPTER 05

INSIDE THE BONE

HOW TO LOOK INSIDE

Bones might seem simple and basic, but they are quite complex and made of many parts and components.

To understand what something looks like on the inside, scientist often remove a section of it to see deeper. This is like slicing an apple to study the seeds.

Let's dive deeper inside the wonderful world of a bone.

PARTS OF THE BONE

If we focus on the leg bone and look inside it, what would we see?

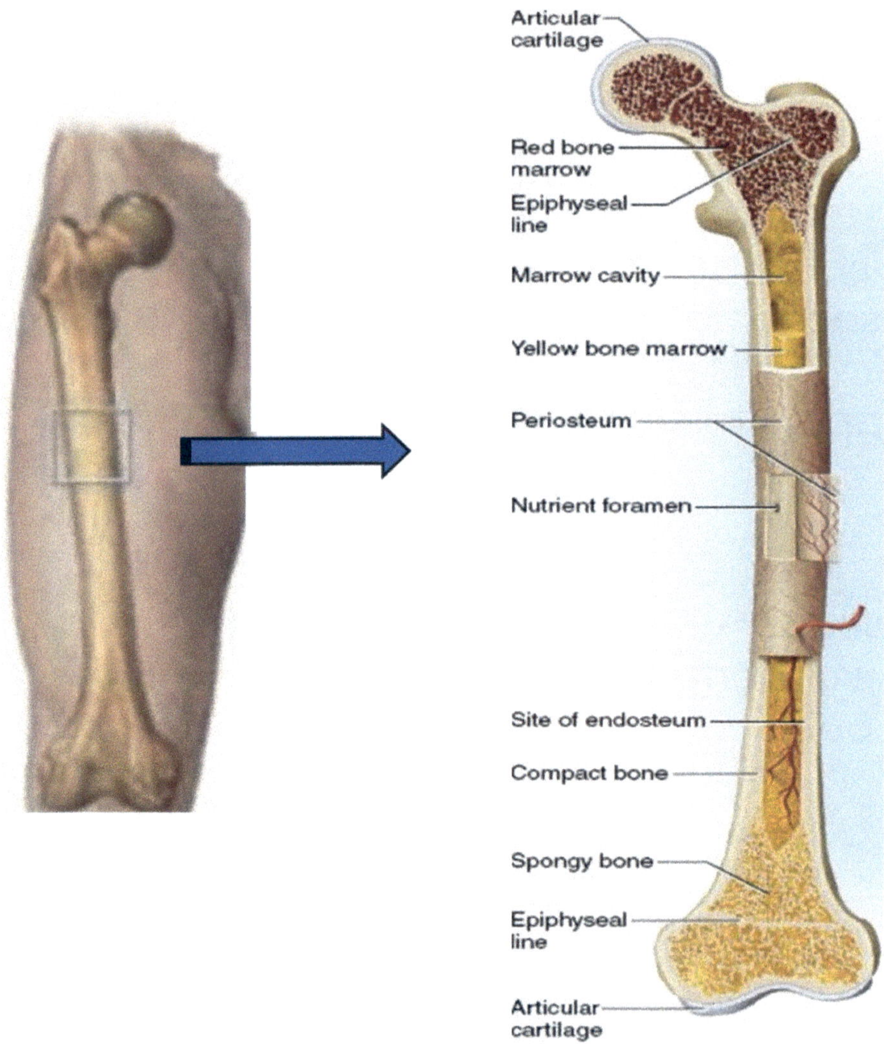

Inside the bone, there are many curious and interesting things.

Let's review some of these parts.

PERIOSTEUM

Bones are made up different parts. The first component we notice is a thin layer that covers the bone and provides protection to the bone. This protective layer is called the **"Periosteum."** The periosteum is made up of small fibers called **"Sharpeys."**

IMPORTANCE

- **Protection**: Acts as a protective layer for bones.
- **Nutrition**: Contains blood vessels that nourish bone tissue.
- **Attachment**: Provides a surface for the attachment of muscles and tendons.
- **Healing**: Plays a vital role in bone repair and regeneration.

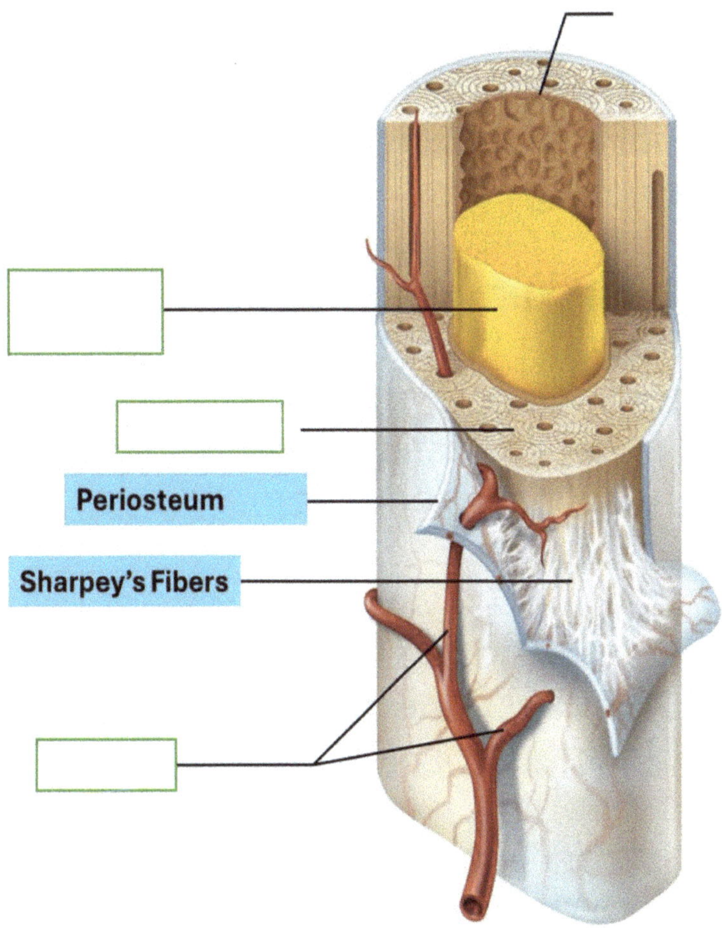

NUTRIENT ARTERIES

The second component is **arteries** that provide nutrients to the bone.

IMPORTANCE

- **Bone Health:** They are essential for delivering nutrients that keep bone tissue healthy and strong.
- **Bone Remodeling:** Support bone remodeling and repair processes by supplying necessary resources.
- **Prevention of Bone** Diseases: Adequate blood supply helps prevent bone diseases like osteoporosis and avascular necrosis.

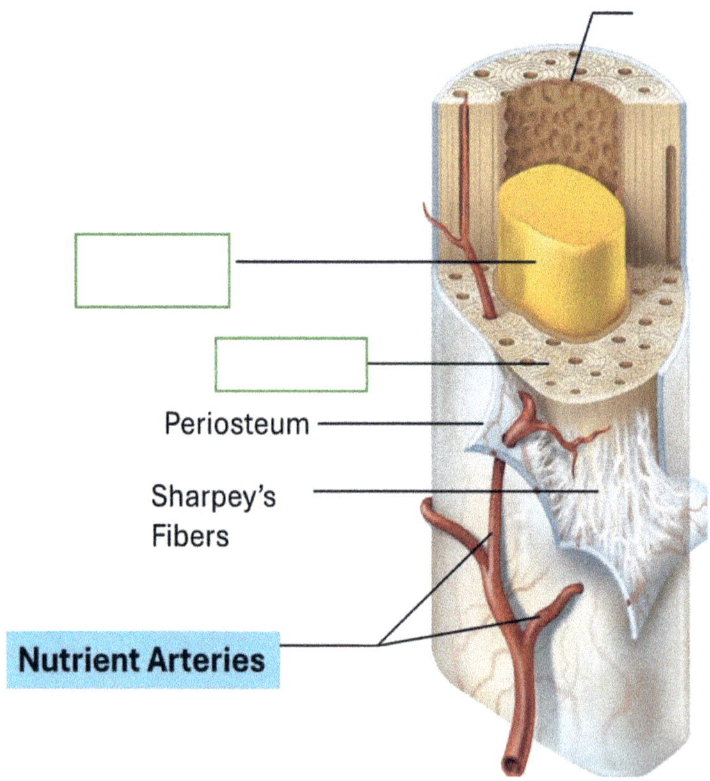

COMPACT BONE

The third part is called the **"Compact Bone."** The compact bone is a dense, hard layer that makes up the outer shell of most bones in the body. It is incredibly strong and provides structural support and protection for the internal components of the bone.

Compact Bone contains microscopic tubes called **Haversian canals**, which house blood vessels and nerves. These canals are arranged in circles around a central canal, forming the Haversian system.

IMPORTANCE

- **Support**: Provide strength and rigidity to bones, allowing them to support the body's weight and withstand mechanical stress.
- **Protection**: Encases and protects the inner spongy bone and bone marrow.
- **Attachment**: Offers a solid surface for the attachment of muscles, tendons, and ligaments.
- **Storage**: Stores essential minerals like calcium and phosphate, which are crucial for various bodily functions.

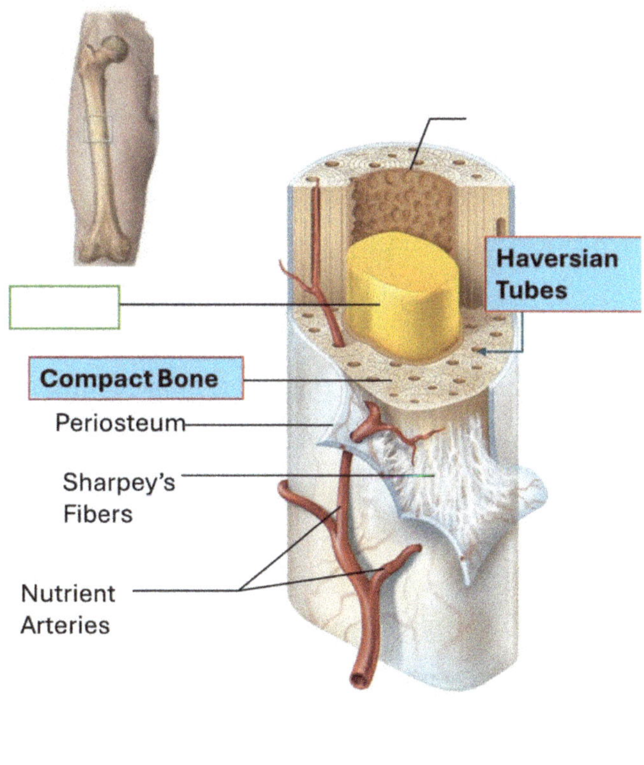

BONE MARROW

The fourth part is called **bone marrow**. Bone Marrow is the soft, fatty tissue inside of the bones in your body. Bone marrow contains cells that produce red and white blood cells and platelets, and it is responsible for making billions of new blood cells each day.

There are two types of bone marrow in your body:
Red: Red bone marrow produces blood cells.
Yellow: Yellow bone marrow stores fat.

IMPORTANCE

- **Hematopoiesis:** The process of producing blood cells.
- **Fat Storage:** Yellow bone marrow stores fat cells, which can be converted into energy if needed.
- **Stem Cells:** Contains stem cells that can differentiate into various types of blood cells, supporting continuous blood cell production.

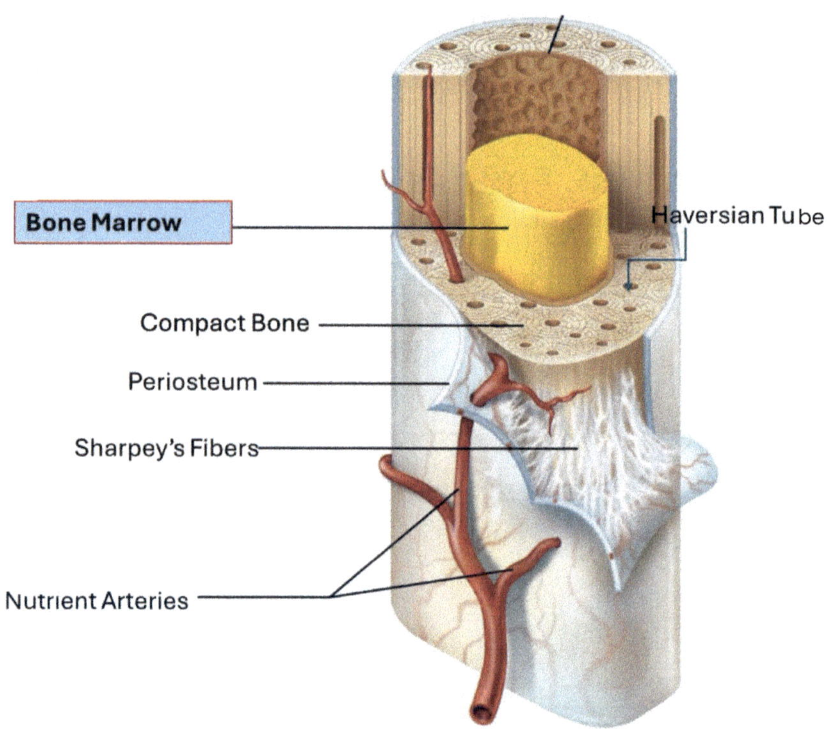

SPONGY BONE

Spongy bone is a type of bone tissue that has a porous or spongy appearance like a sponge you use at home. It is less dense than compact bone and consists of a network of bony parts known as **trabeculae**. It has lots of arteries and contains red bone marrow.

WHY IS IT IMPORTANT?

- **Support and Strength**: The trabecular structure of spongy bone helps distribute the load and stresses placed on bones, making them strong yet light weight.

- **Hematopoiesis:** The red marrow in spongy bone is the primary site for blood cell production (hematopoiesis), generating red blood cells, white blood cells, and platelets.

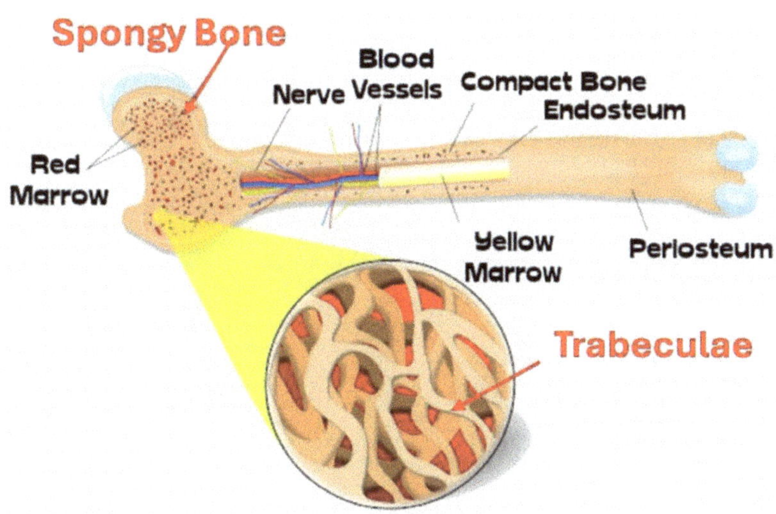

- **Shock Absorption:** The porous nature of spongy bone allows it to absorb shocks and stresses, protecting the bone and the body during activities like walking, running, and jumping.

KNOWLEDGE CHECK

Choose the best answer:

1. The periosteum does what?

Protects the bone | Creates blood

2. Nutrient arteries do what?

Provide nutrients | Keep the bone clean

3. Compact bone is what?

Soft and delicate | Strong and hard

4. The bone marrow creates what?

Blood cells and platelets | Sugar and minerals

CHAPTER 06
BONE GROWTH

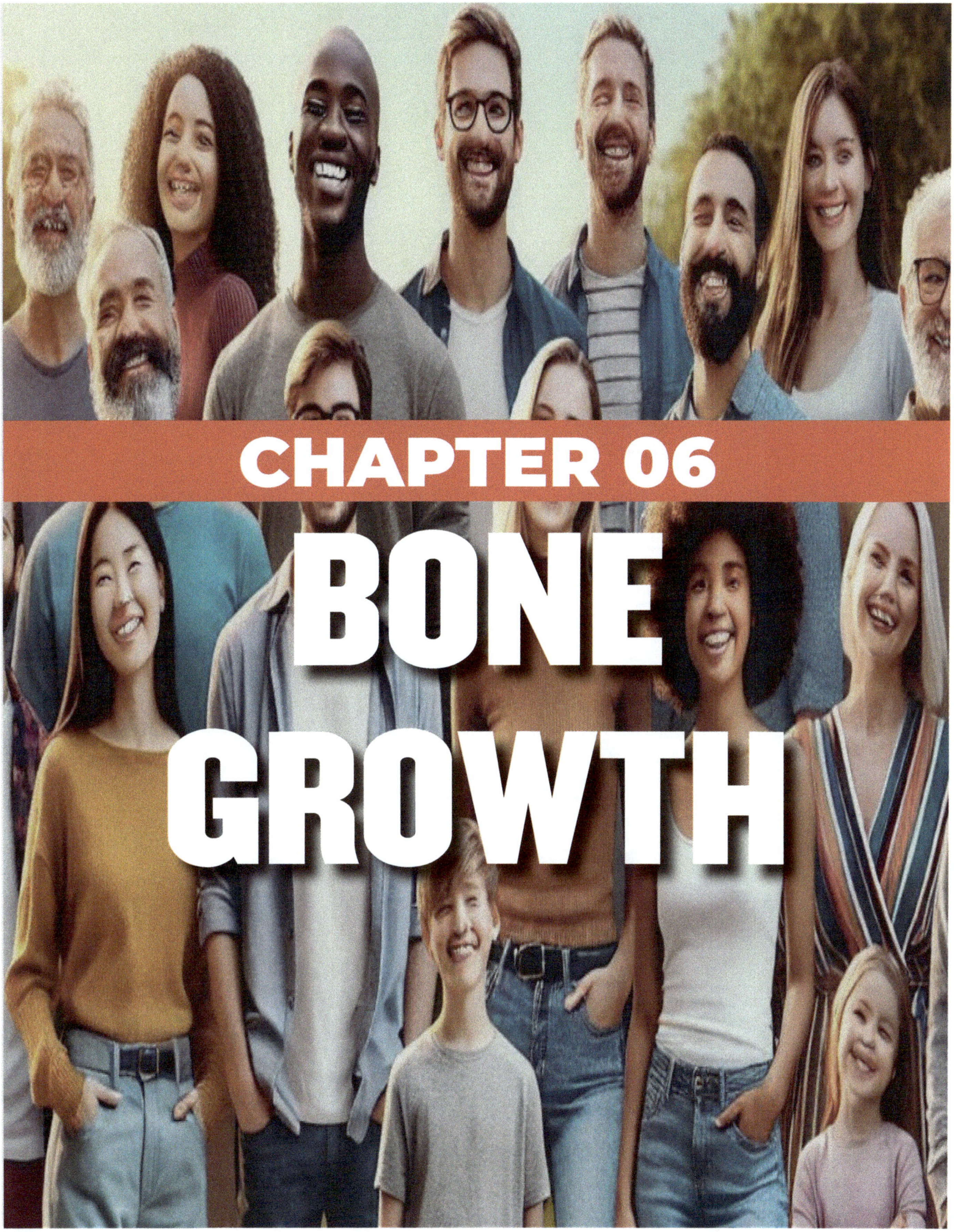

Growing up can be exciting and filled with new experiences. But how do we grow? What causes our bones to become longer?

HOW BONES GROW?

Answer: During the ossification process, the cartilage at the end of bones does not completely transform into bone but continues as cartilage. The remaining cartilage allows the bone to grow. This important area is called "Growth Plates" and if they do not ossify and close, bones can continue to grow and lengthen.

In children and adolescents, bone growth occurs at these **growth plates** (epiphyseal plates) found at the ends of long bones. These plates are made of cartilage and gradually ossify, adding length to the bones.

Growth plates are crucial because they are responsible for the final length of bones and will eventually close once a person reaches adulthood.

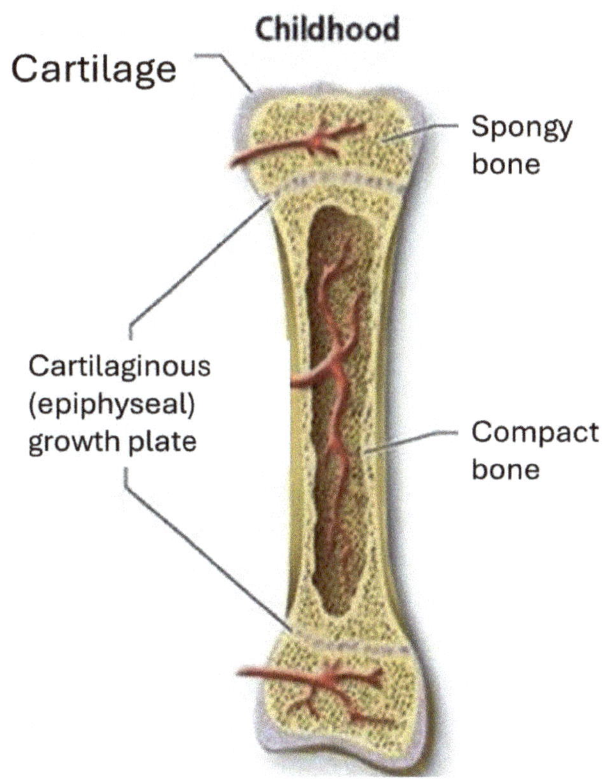

HOW TALL WILL YOU GROW?

Many people want to be tall or taller. But how do you know what your final height will be?

If your growth plate has not formed into solid bone yet, it means there is still some room for growing. There are, however, many methods that can be used to calculate what your height might end up at.

Here is one way:

Your parents' heights play a significant role in determining your own. A general rule of thumb for estimating adult height is to:

1. Add your parents' heights together (in inches or centimeters).

2. Divide that number by 2.

3. Add 5 inches (13 cm) for boys or subtract 5 inches (13 cm) for girls.

WHEN DO GROWTH PLATES CLOSE?

As a general rule:

Girls: Growth plates usually close around ages 13 to 15.
Boys: Growth plates generally close a bit later, around ages 15 to 17.

Factors Influencing Closure

Genetics: Family history can influence when growth plates close.

Hormones: Hormones like growth hormone, estrogen, and testosterone play key roles in the timing of growth plate closure.

Nutrition and Health: Proper nutrition and overall health can impact bone growth and development.

HOW TO TELL WHEN GROWTH PLATES ARE CLOSED

An X-ray can help determine if the growth plate is closed or not. Growth plates appear as dark lines on an X-ray.

If the growth plates are not visible on an X-ray, these bones are done growing. There isn't any more material for the body to grow into new bones. However, not all growth plates in the body close at exactly the same time.

For example, closed growth plates on the wrist does not guarantee that the leg bones are done growing.

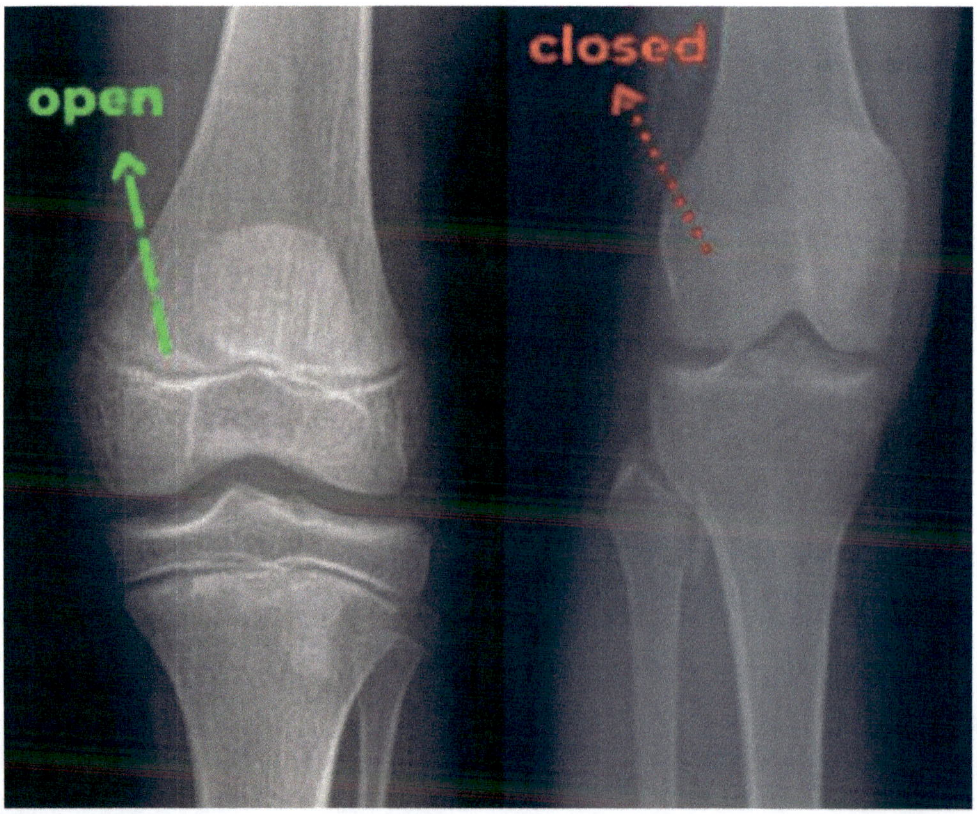

WHAT HELPS BONES TO GROW?

Several nutrients and hormones play key roles in bone growth and health and can help you reach your full genetic potential:

Calcium and Vitamin D: Essential for bone mineralization and strength.

Growth Hormone: Stimulates growth in childhood and adolescence.

Estrogen and Testosterone: Promote bone growth and help in the closure of growth plates during puberty.

Always consult with parents and the family doctor when trying to grow bones. They will provide the best and safest solution.

EXERCISE AND BONES

Exercise especially weight-bearing exercises, stimulates bone growth and remodeling. The mechanical stress applied to bones during exercise signals the body to strengthen the bones.

FUN FACTS

About 10% of our bone structure is replaced each year through a process called bone remodeling!

KNOWLEDGE CHECK

Determine if the statement is true or false.

1. If growth plates do not close, bones can continue to grow. — True or False
2. Each person's growth plate closes at different times. — True or False
3. A person's growth is determined by their family's history. — True or False
4. Good nutrition and exercise can help bones to grow to their fullest potential. — True or False

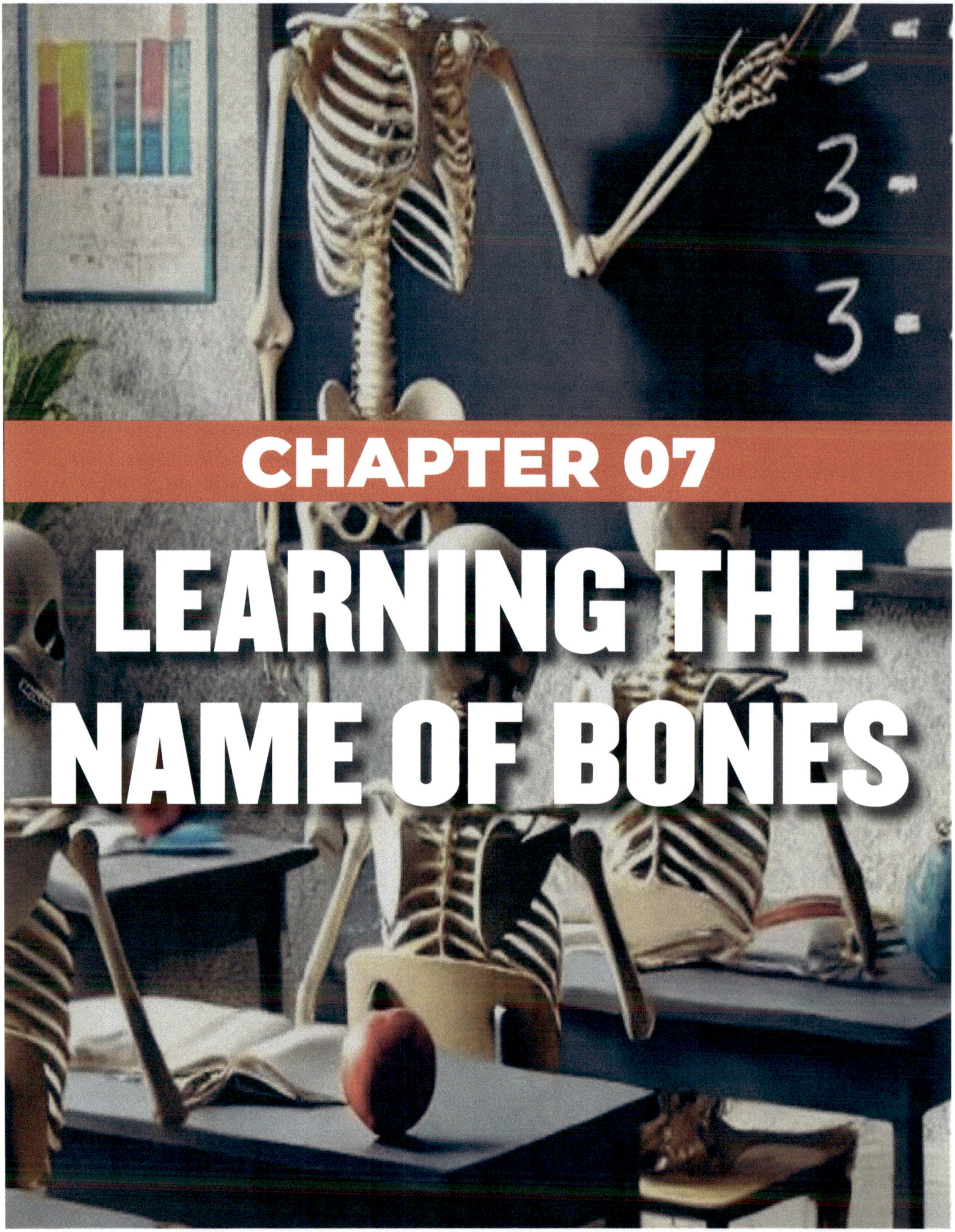

CHAPTER 07

LEARNING THE NAME OF BONES

Learning the name of the bones in our body can be fun and challenging. The best way is through repetition and practice. Start from the top of the head (cranium) and work your way down to the toes (phalanges).

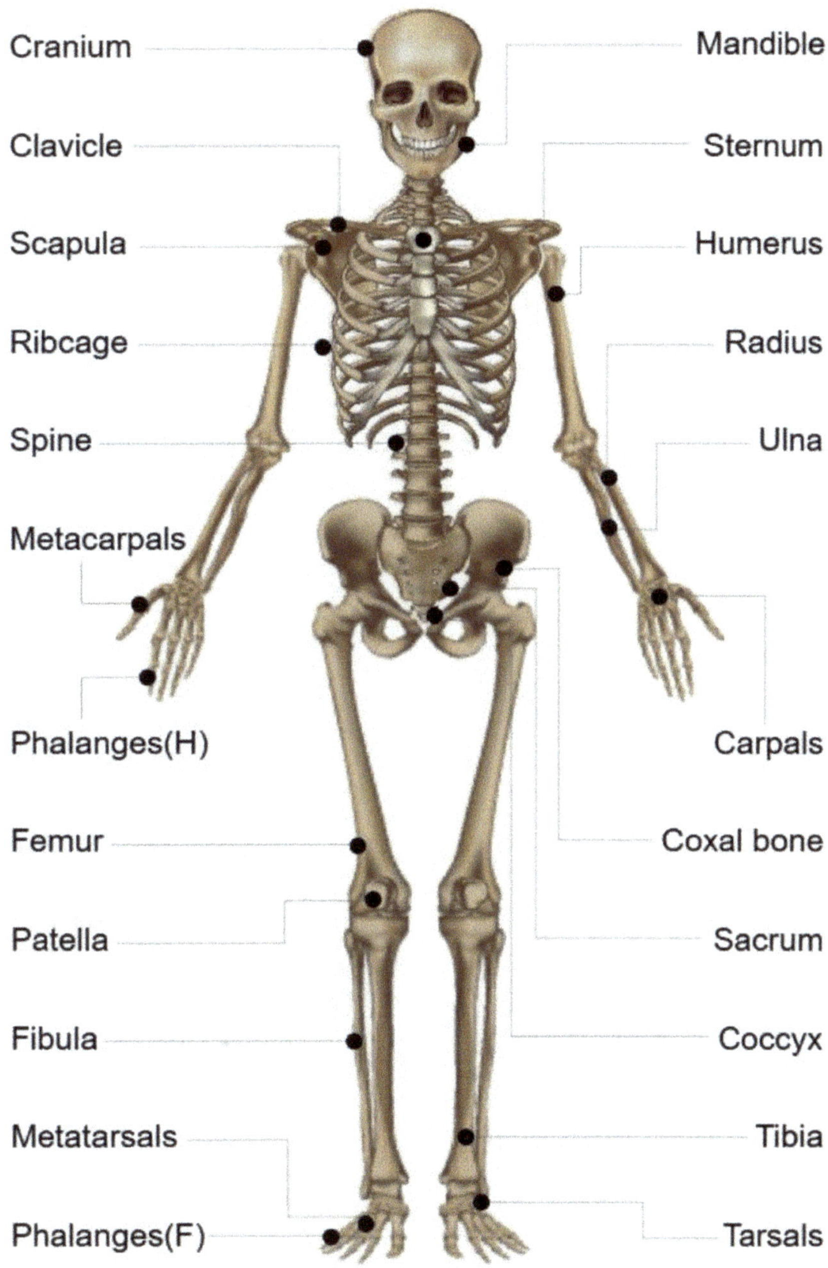

SKULL BONES

The human skull is made up of **22** bones: **8** cranial bones that protect the brain and **14** facial bones that form the structure of the face

Skull bones are connected by immovable joints called **sutures.** These zigzag seams fuse during childhood, contributing to the skull's strength and rigidity.

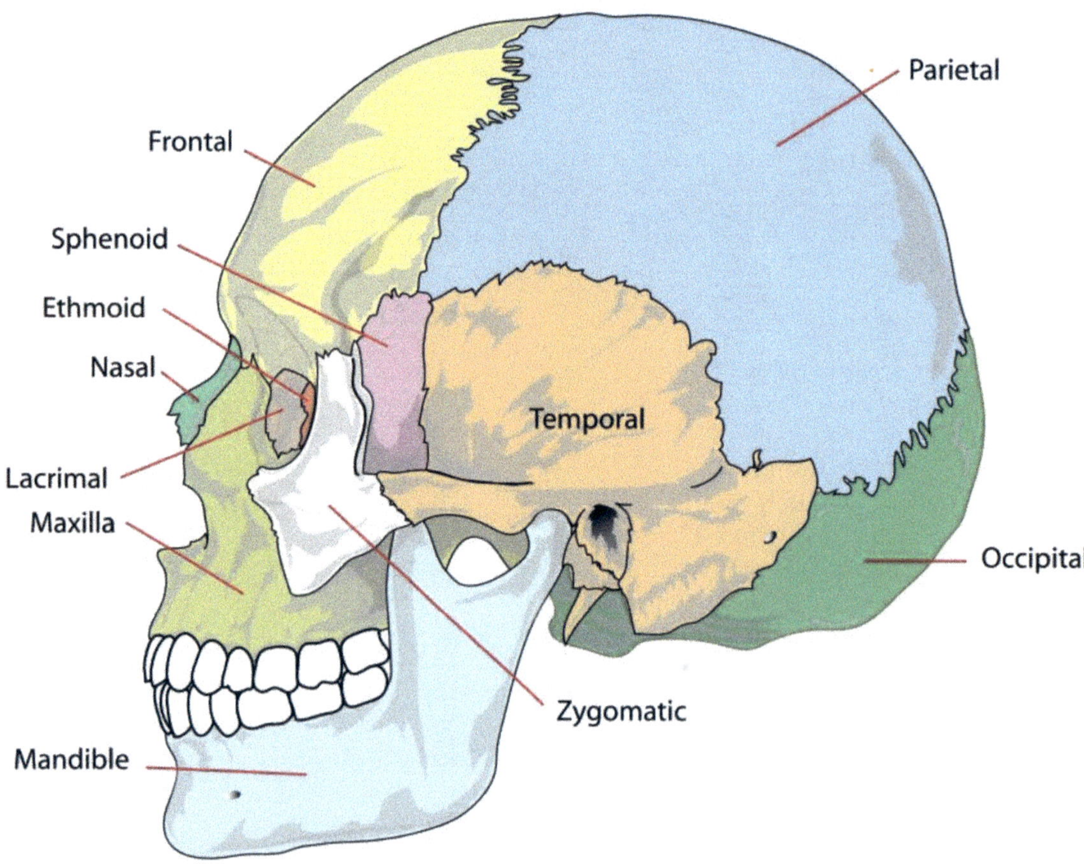

BONES IN THE SPINE

Vertebral Column (spine): 33 bones in children (some fuse as we age, resulting in 24 bones in adults)

The individual bones running along the spine are called **vertebrae**. They are stacked on top of each other with a small soft protective disk between them. The spinal cord runs up through the vertebrae and attaches to the brain.

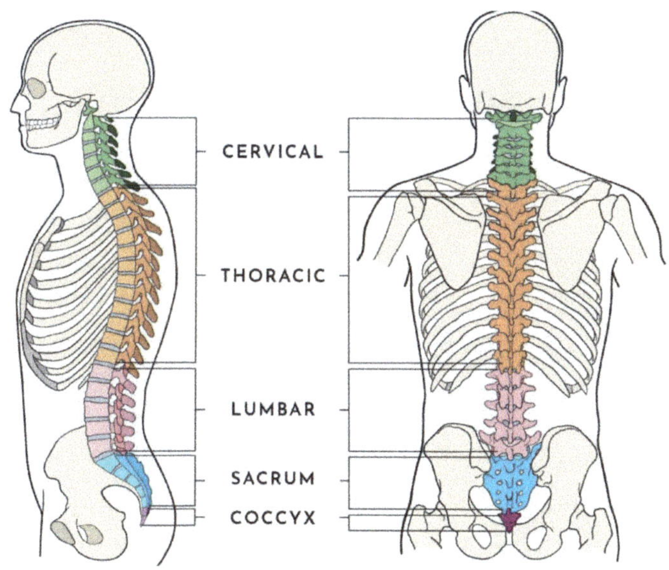

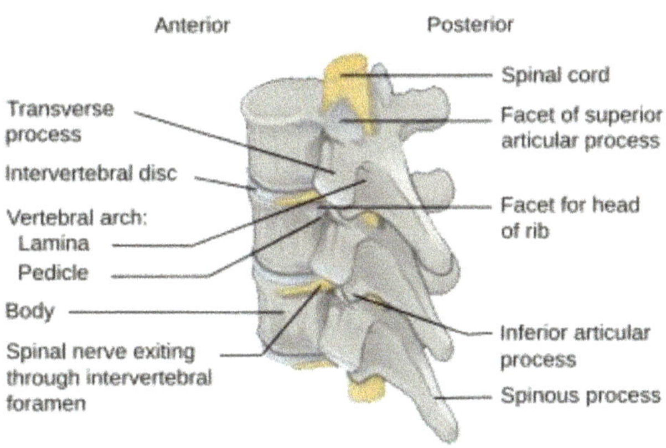

RIB CAGE

Humans typically have 24 ribs, arranged in 12 pairs. However, some people have an extra rib or one fewer, known as rib anomalies.

False and Floating Ribs

•The first seven pairs are called "true ribs" because they attach directly to the **sternum**

•The next three pairs are **"false ribs"** because they attach to the sternum indirectly via cartilage.

•The last two pairs are **"floating ribs"** because they do not attach to the sternum at all.

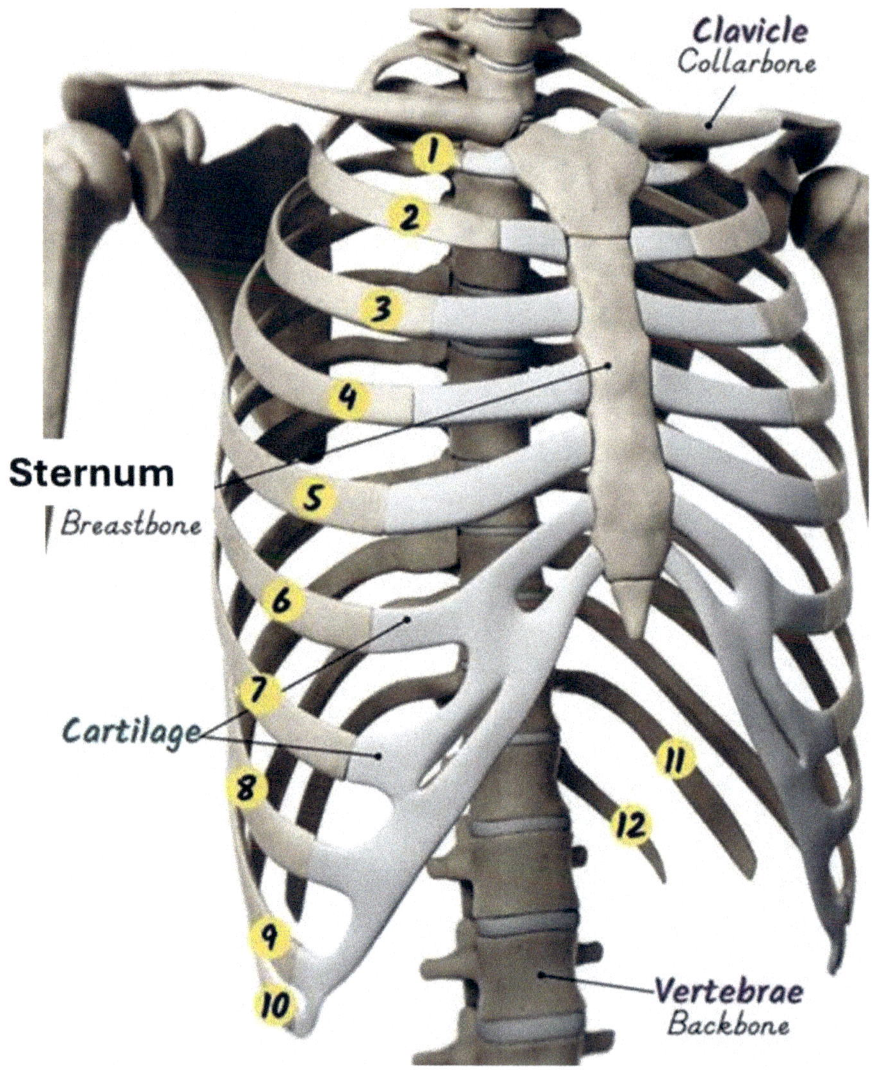

SHOULDER BONES

Shoulder Girdle: 4 bones (2 clavicles, 2 scapulae)

The **clavicle** connects the arm to the body, acting as a strut that supports the shoulder. It is connected to the sternum (breastbone) at one end and the **scapula** (shoulder blade) at the other.

The clavicle is the only long bone in the body that lies horizontally. It has a unique double-curved shape, which increases its strength and helps to absorb shock.

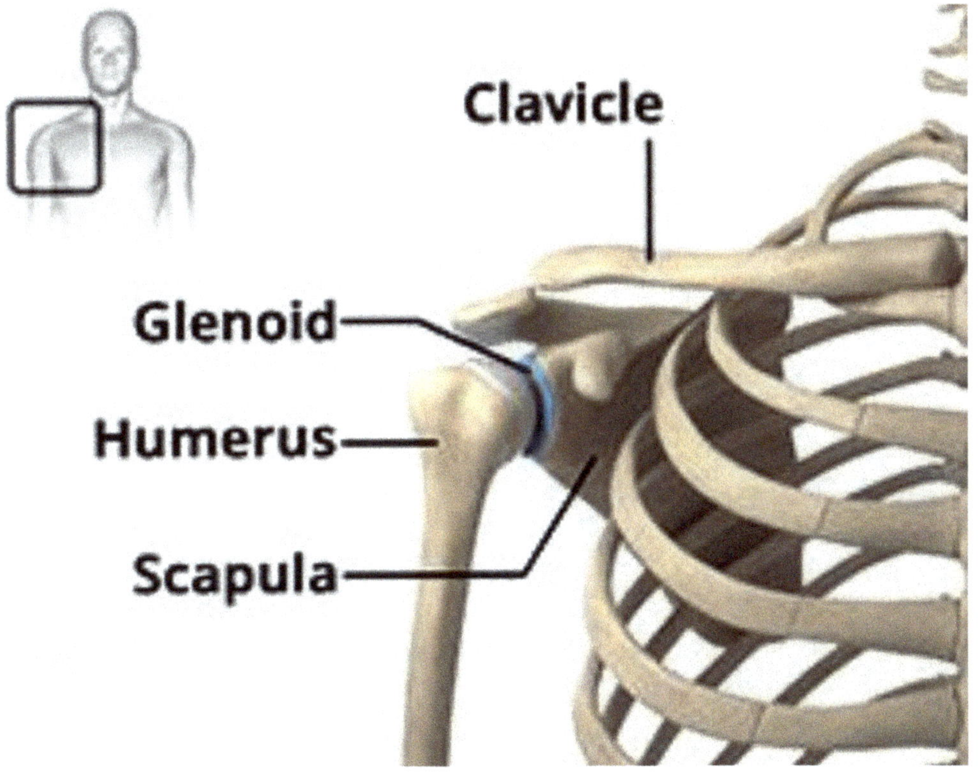

ARM AND HAND BONES

1. Three Main Bones
The arm consists of three main bones: the humerus in the upper arm, and the radius and ulna in the forearm.

2. Humerus
The humerus is the longest bone in the arm and runs from the shoulder to the elbow. It gets its name from the Latin word for "shoulder."

3. Radius and Ulna
The radius and ulna are the two bones in the forearm. The radius is located on the thumb side, and the ulna is on the pinky side.

Did you know?

The "funny bone" isn't actually a bone. It's a nickname for the ulnar nerve that runs along the inside of the elbow. When you hit it, you feel a tingling sensation that can be quite uncomfortable.

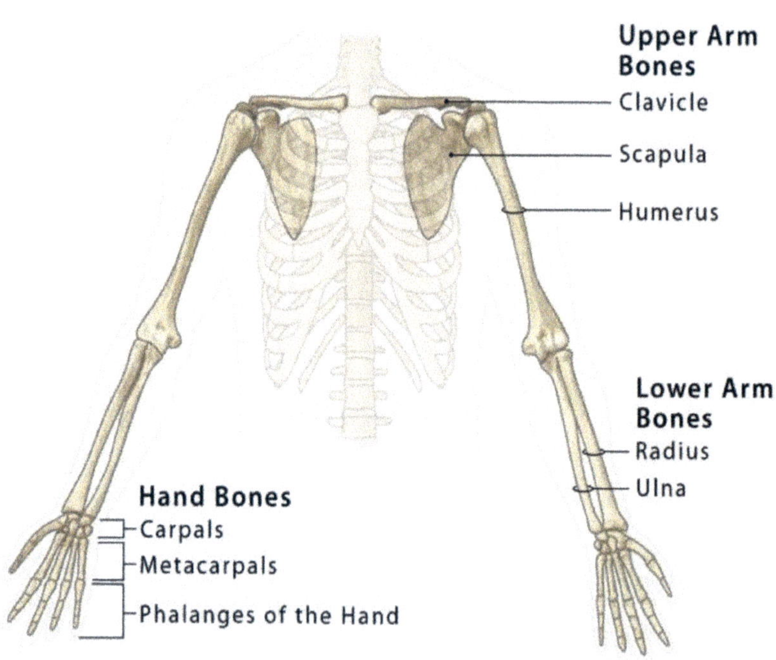

HAND BONES

1. Design
Each hand has 27 bones: 8 in the wrist (carpals), 5 in the palm (metacarpals), and 14 in the fingers (phalanges).

2. Unique Patterns
The bones of the hand are arranged in a way that allows for a wide range of movements, enabling complex tasks like writing, playing musical instruments, and using tools.

3. Moveability (dexterity)
The thumb has a unique saddle joint that allows it to move across the palm, enabling the grip and handling of objects. This is called opposable thumb, a trait crucial for human dexterity.

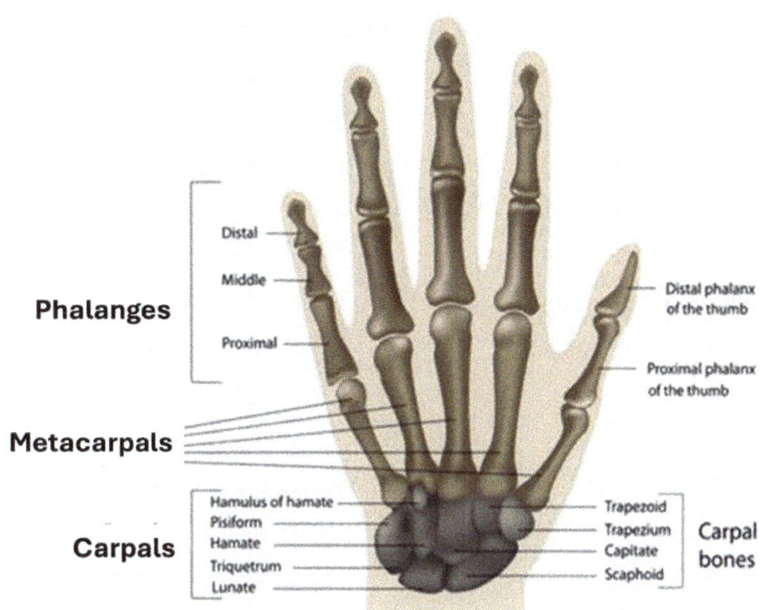

HIP BONES

Pelvic Girdle: 2 bones (hip bones)

Each hip bone is made up of three bones: the ilium, ischium, and pubis. These bones fuse together during adolescence to form a single, solid bone.

The hip joint, where the femur (thigh bone) meets the pelvis, is one of the strongest and most stable joints in the human body. This allows for a wide range of motion while supporting the body's weight.

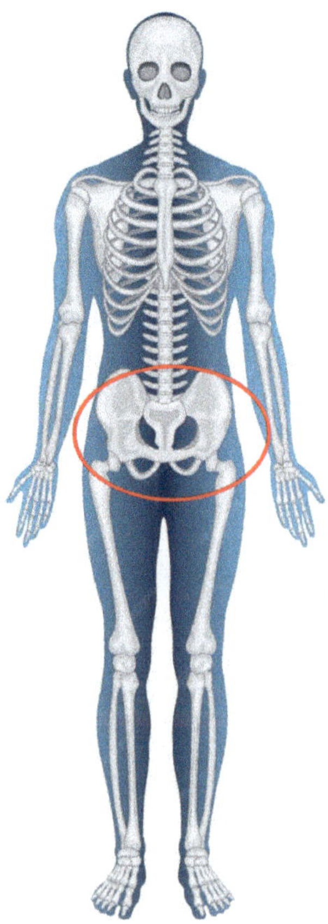

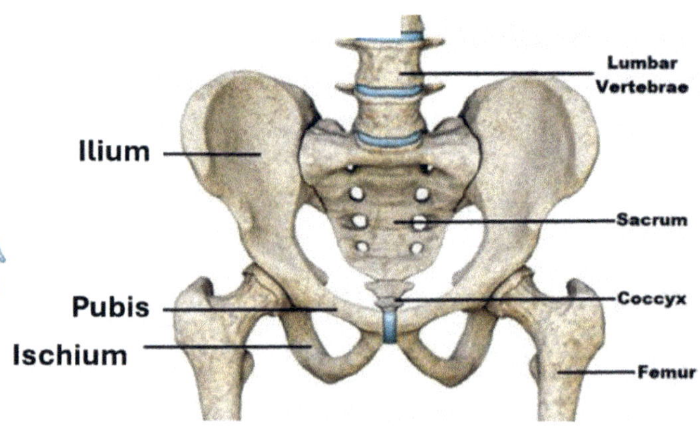

LEGS AND FEET BONES

Legs and Feet: 60 bones (30 in each leg)

1.Longest Bone
The **femur**, or thigh bone, is the longest and strongest bone in the human body. It can support up to 30 times the weight of an adult.

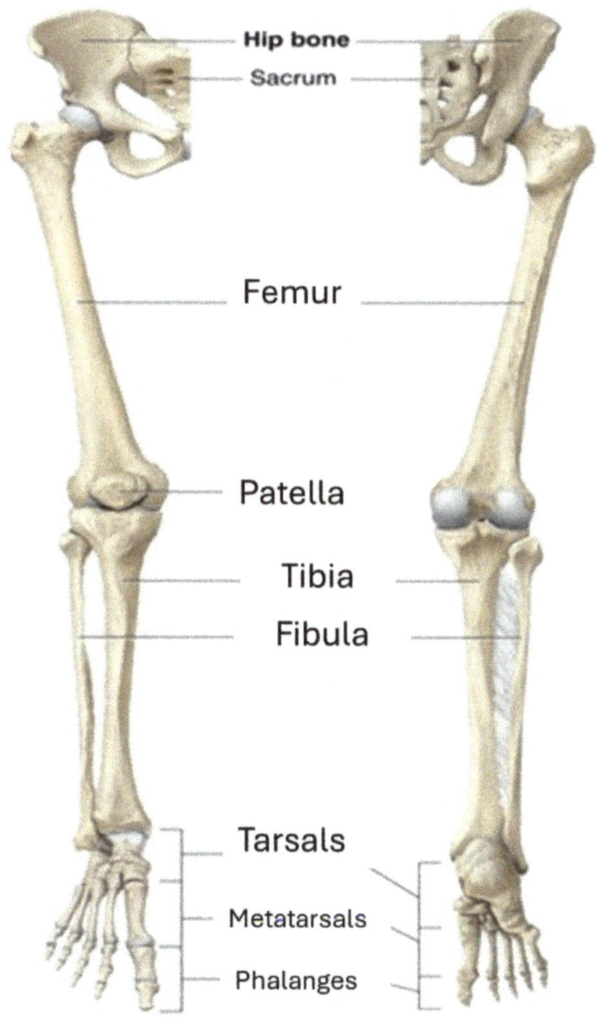

2. Kneecap
The **patella**, or kneecap, is a small, triangular bone that protects the knee joint and enhances the leverage of the thigh muscles.

3. Two Lower Leg Bones
The lower leg has two bones: the **tibia** (shinbone)and the **fibula**. The tibia is larger and bears most of the weight, while the fibula provides stability and muscle attachment.

FEET BONES

1. Bone Count
Each foot contains 26 bones, making up a total of 52 bones in both feet. That's about a quarter of all the bones in the human body!

2. Design
The bones in the feet are divided into three groups: **tarsals** (7 bones), **metatarsals** (5 bones), and **phalanges** (14 bones).

3. Largest Foot Bone
The calcaneus, or heel bone, is the largest bone in the foot and plays a key role in supporting body weight.

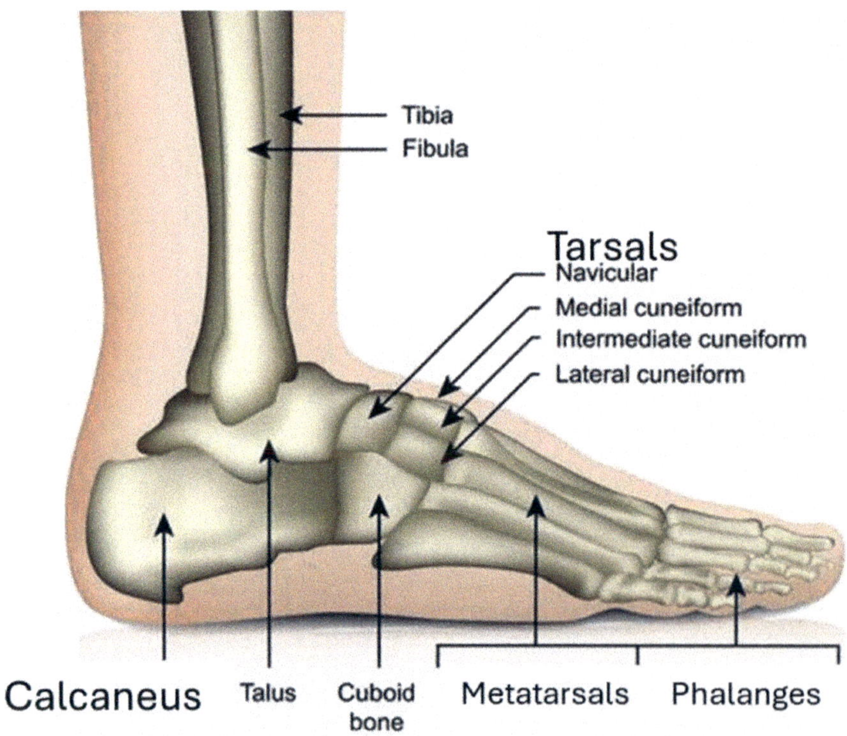

FILL IN THE BLANKS

See if you can name the major bones in this picture.

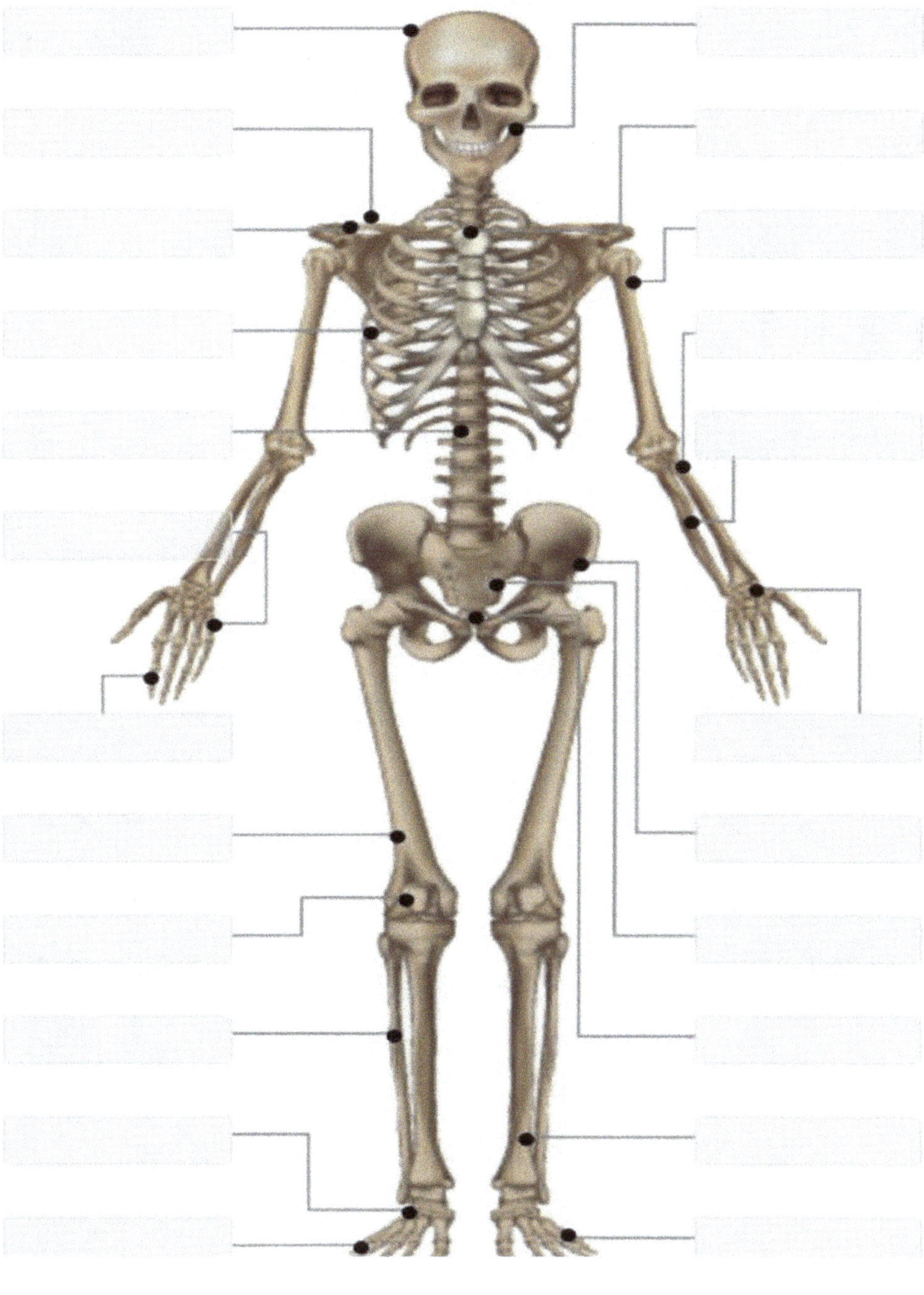

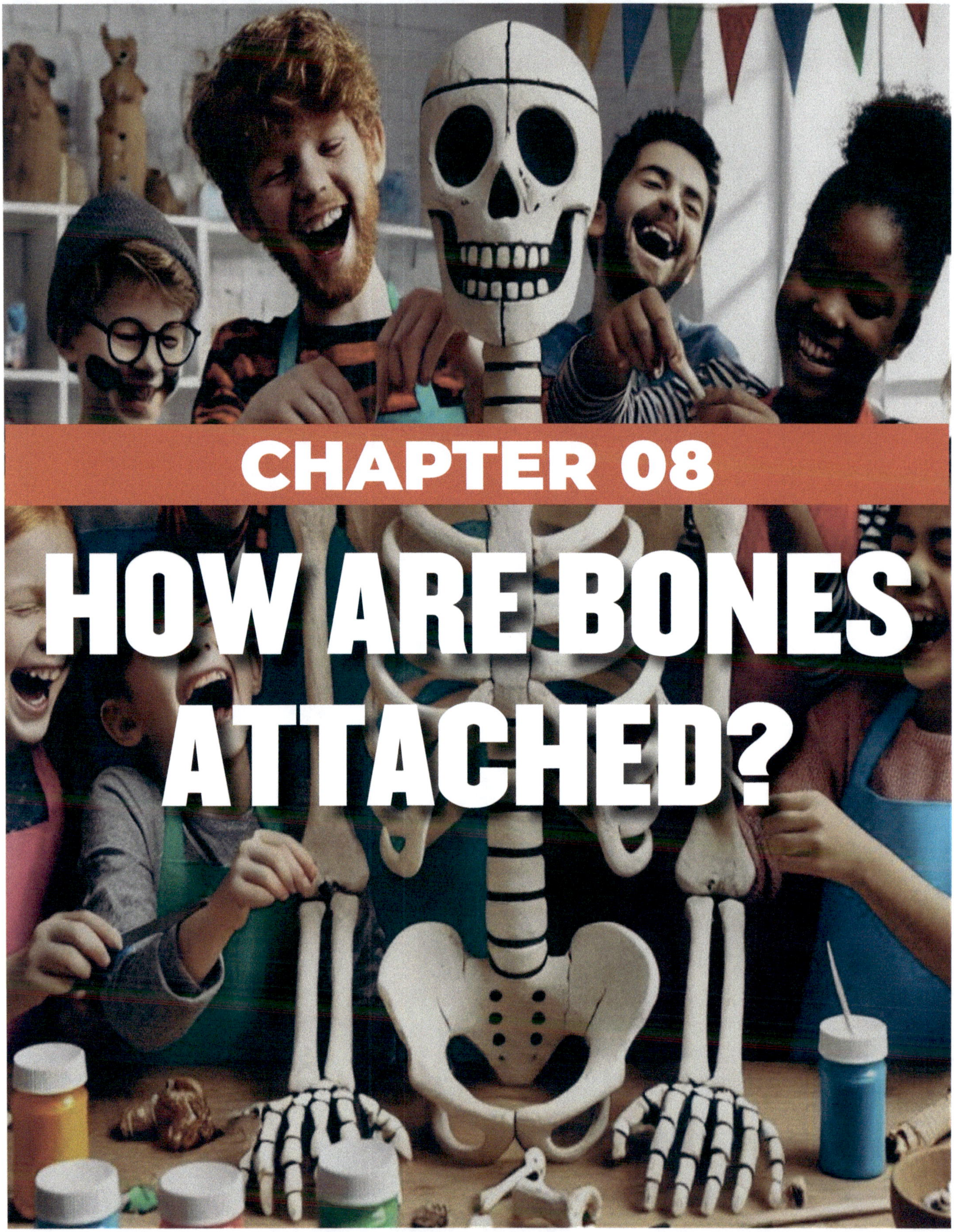

CHAPTER 08

HOW ARE BONES ATTACHED?

HOW ARE BONES ATTACHED?

There are 206 bones in the body. But how are they attached together? Can you guess the correct answer?

- Glue
- Strings
- Staples
- Ligaments

LIGAMENTS

Ligaments are strong, fibrous tissues that connect bones to other bones. They provide stability to joints by holding the bones together.

Example: The anterior cruciate ligament (ACL) in the knee helps keep the bones of the knee joint in place. A joint is the location where two bones meet.

Think of ligaments like strong, stretchy tapes that connect your bones together at your joints, like your knees and elbows.

A joint is the location where two bones meet.

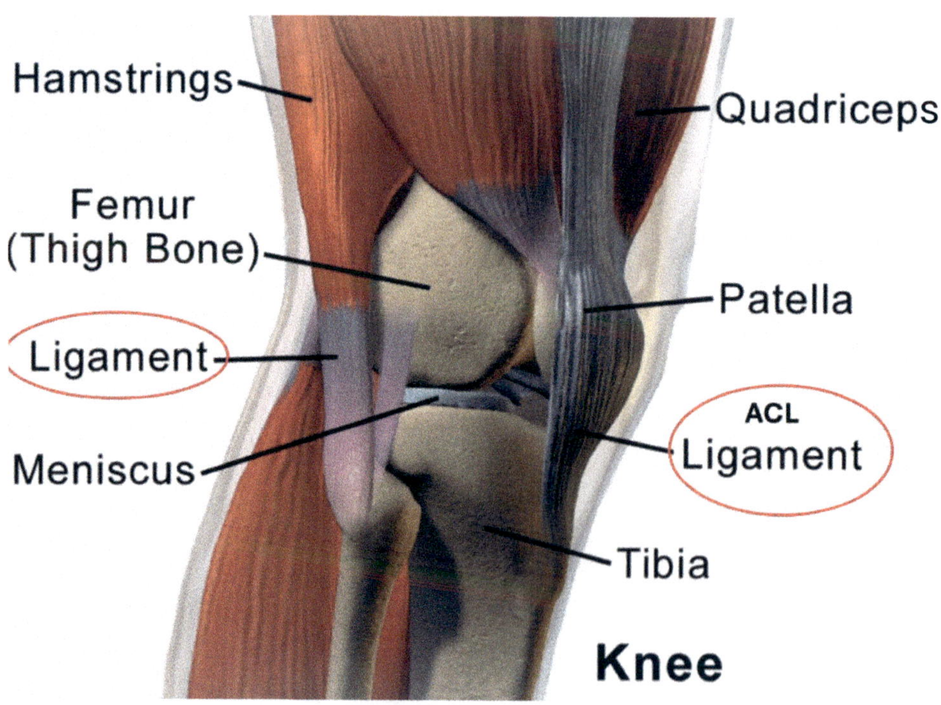

TENDONS

Tendons are tissues that connect muscles to bones. While they don't directly connect bones to each other, they play a crucial role in movement and stability by transmitting forces from muscles to bones.

Example: The Achilles tendon connects the calf muscles to the heel bone, allowing for movements like walking and running.

Tendons are like tough ropes that attach muscles to bones, helping you move your arms and legs.

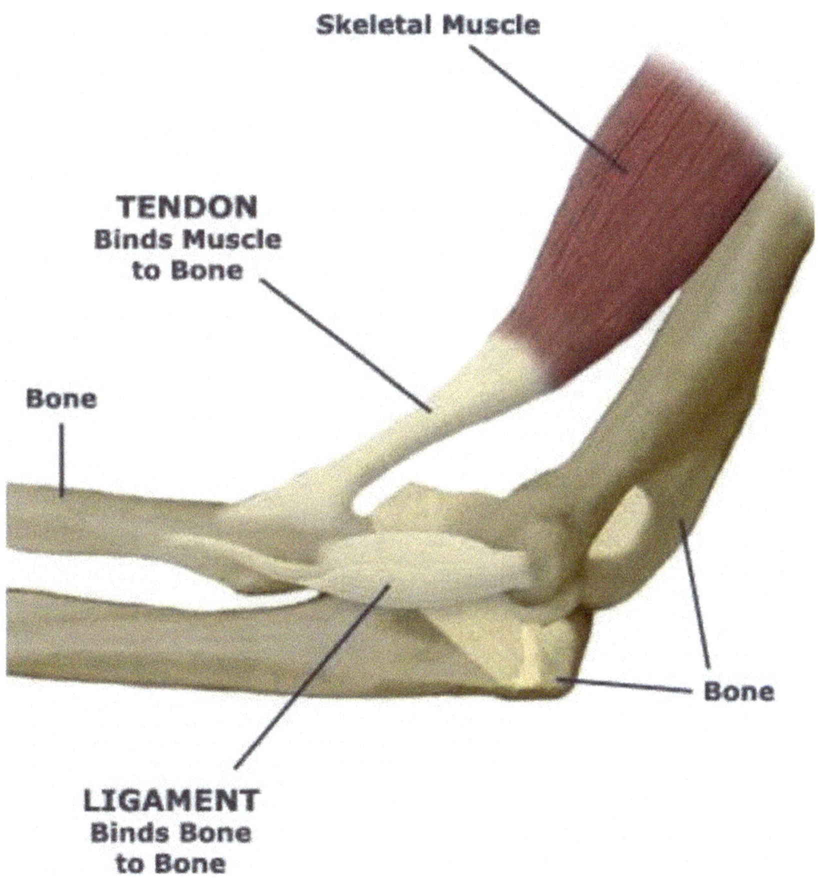

JOINT CAPSULES

Joint capsules are connective tissue structures that surround joints. They contain synovial fluid, which lubricates the joint and helps keep the bones together.

Example: The shoulder joint capsule encloses the shoulder joint, providing stability and allowing for a wide range of motion.

A connective tissue is one of the four main types of tissue in the human body. It plays a critical role in providing support, and cohesion to other tissues and organs throughout the body. Think of it like the special pieces that hold the LEGO blocks together and make everything strong and flexible.

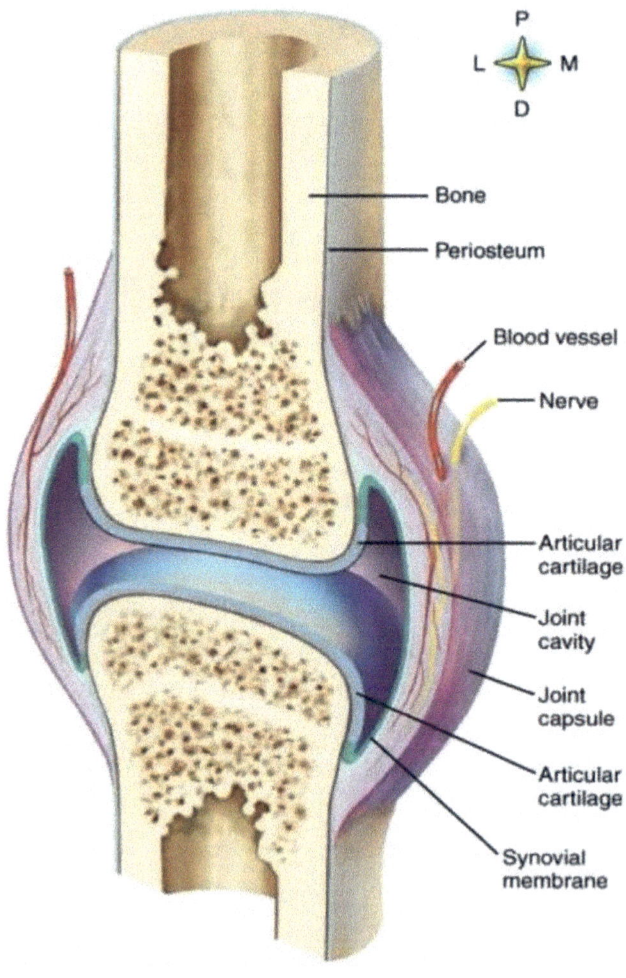

MUSCLE

Muscles surrounding the bones and joints provide additional support and stability. They help hold the bones in place through the tension they create.

Example: The muscles of the rotator cuff in the shoulder help stabilize the shoulder joint.

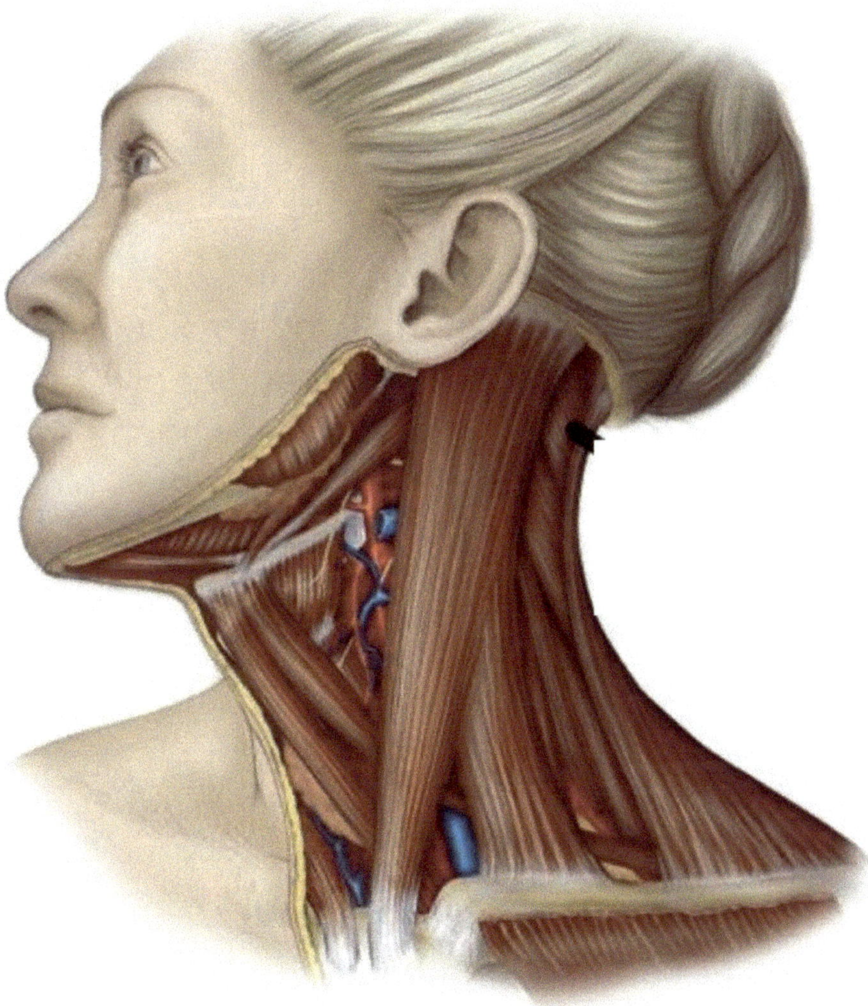

BONE SHAPE AND FIT

The shape and fit of bones at a joint can help keep them together too. For example, the ball-and-socket design of the hip and shoulder joints allows for stability and flexibility.

Example: The socket of the pelvis snugly holds the ball of the femur, creating a stable yet movable hip joint.

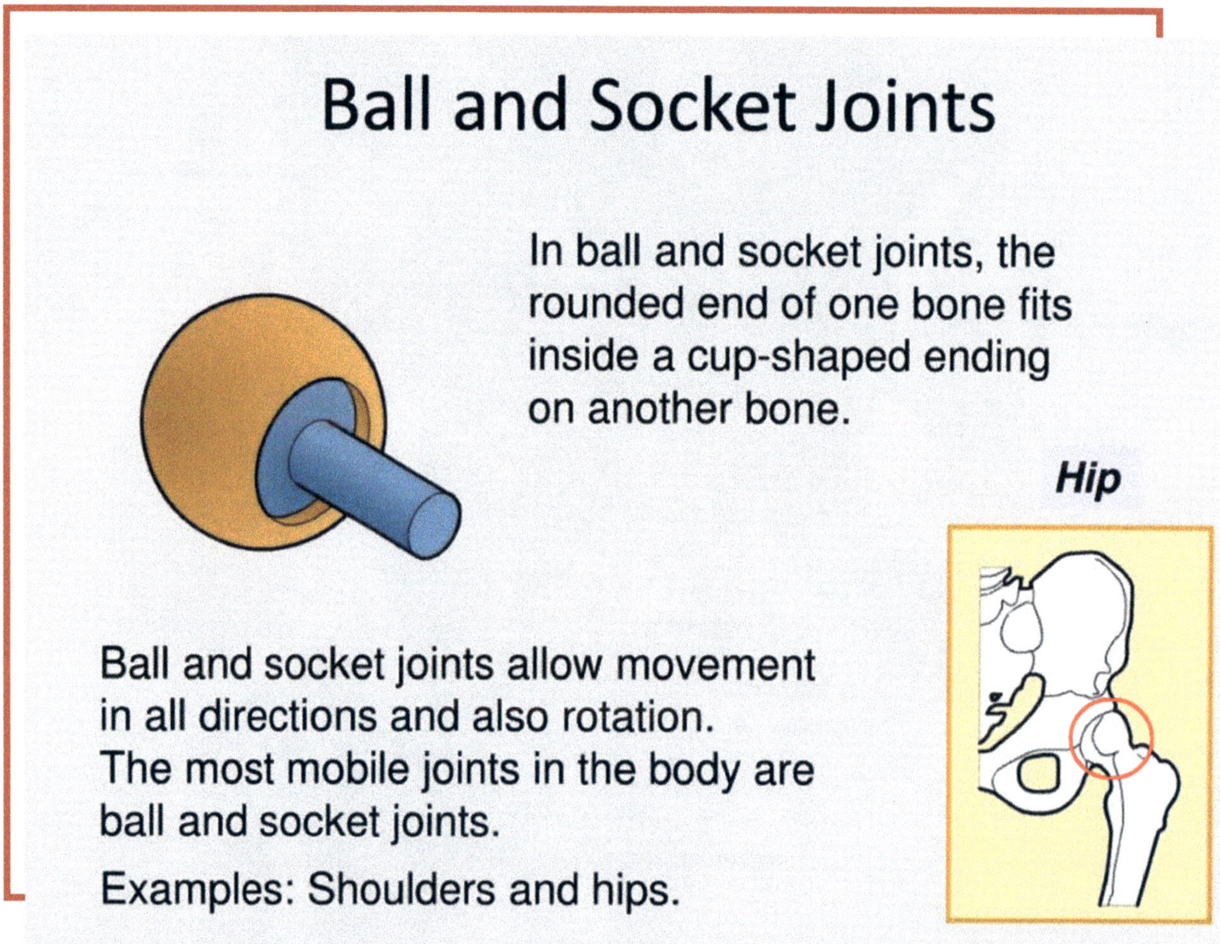

KNOWLEDGE CHECK

1. *How many bones are there in the human body?*

2. *How are bones attached to each other?*

3. *What determines how tall a person will grow?*

4. *How do bones grow?*

5. *What is a tendon?*

CONCLUSION

I hope you have enjoyed reading this book. We learned a lot about bones and the human skeleton. Our bones are crucial, beautiful and strong. They can last for very long periods of time. Precious blood is created inside the bone and our bones store important nutrients.

Even though learning the name of all 206 bones in our bodies might seem like a lot to memorize, with practice and time you will be able to recite them by heart.

Stay safe and never stop learning!

AUTHOR'S BIO

The author is an educator and a passionate lover of children. He received his Bachelor of Arts in Biology from Edgewood College in Madison, Wisconsin and a second B.A. in Healthcare Management from Southern Illinois University Carbondale where he also completed his M.A in Engineering. He has a wealth of valuable experience dealing with children of varying age ranges. The author does not only love children, but he understands that when they have the best form of education, it has a far-reaching effect on the wider society. As a result, he is on a mission to make sure that children learn as much as they can.

REFERENCES

How Many Bones In The Human body. Anne Helmenstine. April 6, 2024 https://sciencenotes.org/how-manybones-are-in-the human-body/

The Skeletal System. https://www.theskeletalsystem.net/skull-bones

Gray's Anatomy for Students. Richard L. Drake. March 24,2023. Elsevier, 5th edition

Growth Plates. Nemours Kids Health. https://kidshealth.org/en/parents/growth-plates.html

How To Know If Your Child's Growth Plates Are Closed. Elisa Cinelli. March 27, 2024. https://www.familyeducation.com/teens/puberty-sex/how-to-know-if-your-childs-growth-plates-are-closed

Dorland's Illustrated Medical Dictionary. Dorland. Dec 30, 2019. Elsevier (33rd ed.)

Bones. https://www.ck12.org/biology/bones/lesson/Bones-Advanced-BIO-ADV/

Interior images by Shutterstock, Microsoft Stock Images, iStock, and Zach's Creative Space.

www.ingramcontent.com/pod-product-compliance
Lightning Source LLC
LaVergne TN
LVHW070857210926
848482LV00016B/959

* 9 7 9 8 9 9 7 3 6 6 1 2 4 *